T0325877

Success in Innovation

Success in Innovation

Improving the Odds by Understanding the Factors for Unsuccess

Jan Verloop

ELSEVIER

AMSTERDAM • BOSTON • HEIDELBERG • LONDON • NEW YORK • OXFORD
PARIS • SAN DIEGO • SAN FRANCISCO • SINGAPORE • SYDNEY • TOKYO

Elsevier
30 Corporate Drive, Suite 400, Burlington, MA 01803, USA
525 B Street, Suite 1800, San Diego, CA 92101-4495, USA

First edition **2013**

Library of Congress Cataloging-in-Publication Data
Verloop, Jan. Success in Innovation – Improving the Odds by Understanding the Factors for Unsuccess – First Edition.
pages cm
Includes index.
ISBN 978-0-12-397889-9
1. Technological innovations–Economic aspects. 2. Industrial management. 3. Business enterprises–Finance. 4. Success in business. I. Title.
HD45.V37 2013
658.4'063–dc23

2012051676

British Library Cataloguing in Publication Data
A catalogue record for this book is available from the British Library

Printed and bound in USA
13 14 15 16 17 10 9 8 7 6 5 4 3 2 1

ISBN: 978-0-12-397889-9

Cover design: Robert J. Verloop

This book has been manufactured using Print On Demand technology. Each copy is produced to order and is limited to black ink. The online version of this book will show color figures where appropriate.

Contents

Foreword

The rationale for this book is based on a hard fact of life that for every successful innovation, there are ten failures. However, a quick trawl through the innovation literature would probably leave you thinking that innovation was all about finding the secrets of success, the common 'magic bullets' that characterize successful innovations. Small wonder then that many readers, like myself, are left wondering how can they learn from these success factors, how would they apply to me? The problem is that we learn more from mistakes and errors than we ever can from success. This might be why so many of the 'magic bullets' in the literature seem so difficult to apply and leave a sense of dissatisfaction.

Jan Verloop is a brave author; he has thought a great deal about innovation and taken this point about learning from mistakes fully on board. Building on his earlier work Insight in Innovation, itself a brave approach seeking to understand the laws of innovation, this book focuses on not only why innovations succeed but also why they fail. I think this is an important insight; we can learn so much more this way. It is harder to write about errors, mistakes, and failure, especially when examples are required, but Jan takes this on squarely, deploying powerful examples to reinforce key points.

He does not entirely dismiss the positive side, however, but does manipulate the terminology a little, so his key unsuccess factors are learning points for avoiding failure. He also makes the important point that stopping the innovation process at an early but appropriate stage for the right reasons can be seen as a success, not a failure. It is important for anyone managing innovation to understand this key point. Jan made this point to me many years ago when he was managing a highly innovative research portfolio. In that quirky Dutch way, he explained, toying with negatives, that the crucial factor in managing the research portfolio was that *"I have to keep my ability to kill projects alive."* When I deciphered the point, it was a profoundly important insight that culling the weaker projects early kept the whole portfolio innovative and healthy. This was a far cry from other research portfolios I have seen, especially in government research, where many projects take on a life of their own and are often impossible to cull. A surfeit of these 'zombie projects' then drag the rest of the portfolio down into the mundane, killing off any chance of innovation.

This book will also provide a good mental framework for thinking about innovation, but it is practical, not academic. Jan will explain how difficult it is to attempt to refinance an innovation while it is crossing the Valley of Death, and why this is so. Using the analogy of a quest, this important book on innovation will give the reader a touchstone and insight on common traps to avoid.

A friend of mine who is teaching mainly postgraduate students at Cambridge asked me earlier this year for a recommendation. "*What single book should I recommend to these bright students, the one book they should read before they start out on a career starting up high technology innovative companies?*" At the time I could not answer him; there are many books on the theory of innovation and entrepreneurship, but they are mostly too academic. There are also many practical books propounding the 'magic bullet' success factors of innovation, but they are too hard to apply. I think I can now answer that question: learn from other's mistakes in a practical and positive way by reading Success in Innovation.

Steven Bowns
Director, Technology Futures

Acknowledgments

This book is based on the experience from the innovation projects I have been involved in one way or another over the past 15 years. Experience is a critical factor for success and that is why this book contains many case studies. However, case studies about successes are numerous and easy to find, but case studies about failure are fairly rare. There are case studies from large companies that became world news, and I have used a number in the text, but case studies about failures by start-ups are more or less non-existent.

Therefore, I am very grateful to the people who have helped me by finding and writing these stories. For obvious reasons, the stories about recent unsuccess are anonymous, and I can only mention the names of the authors and not link them to specific stories. Steven Bowns, Michiel Cramwinckel, Ernst Horwitz, Ashok Jain, Michael Kesztenbaum, Susan Robertson, Pascal Verloop, and Peter Vermont provided valuable comments and compelling case studies.

Introduction

Starting a business is relatively simple; making a success of a business is a challenge. Creating a novel innovative idea is challenging some of the time, but bringing an innovative product successfully to the market fails most of the time. Success in innovation does not come easy. Failure is typical for innovation; success is only for one out of ten innovation efforts. With such a poor track record, it should be surprising how popular innovation is. Politicians see it as the only way forward for economic growth and consider it as the essential ingredient for nurturing the wealth of the nation. Business people voice similar words about maintaining competitiveness in the international arena. Young starters see innovation as a way to realize their dreams and create their own destiny.

This raises the question whether there is a conflict between belief and reality in innovation that is not recognized or appreciated. Articles and books about innovation focus on the successes in innovation, and these stories are exciting and appealing. But the fact remains that for each success there are many failures, and although at the end the success of the one big win more than compensates the loss of the many, this is of little comfort when one's innovation is one of the many.

This book has a focus on unsuccess in innovation. Most innovations fail; some of them were bound to fail but others could have been a success if only certain things had been done differently. The word unsuccess is used for those innovation efforts that could have been a success but failed as a result of not understanding or mismanaging the innovation process. Success in innovation should not so much be measured in terms of more innovation, but in terms of better innovation: innovation with more chance of success. Success should not be measured in terms of higher research budgets or number of patents, but by the impact of innovation and the increase of the survival rate from idea to a successful business.

Success in innovation is about improving the odds and reducing the risks by avoiding the factors for unsuccess, and this has to be done on a sound understanding of the innovation process. Many entrepreneurs embark on innovation initiatives without fully understanding what has to be done and which hurdles may be in the way to success. The excitement about the new idea tends to overtake the rational analysis of what needs to be done for achieving success.

The book is written from the perspective of the techno-starter that wants to commercialize his technology-based idea. Technology-enabled innovation is the most common type of innovation, but with a bit of adaptation, the concepts and insights can be applied more or less to all types of innovation. Also the word 'product' will be used in the generic sense for the outcome of the innovation, whether the product is a new physical item, process or service, or a 'solution.' This book is also geared to radical innovation with genuinely novel products as distinct from incremental innovation with improved products.

The book is divided into three parts. Part I provides the understanding of the innovation process that every innovator should have in his mental baggage. It describes a generic method for successfully executing an innovation project that has been tested in practice. However, there is no single recipe for success in innovation and each innovator has to find his/her own way, but understanding the logic of the proposed way forward will support finding the right path to success.

Part II examines the typical, most common pitfalls in the innovation process that lead to failure and indicates how these Key Unsuccess Factors can be avoided.

Part III discusses a number of topics that are part of the innovation domain and can be used in support of the innovation effort. They are important potential factors for success, but they can have negative side effects and these are specifically highlighted, as appropriate for a book about avoiding unsuccess.

Because the focus of the book is on factors for unsuccess, it may read a bit negative, but the key message is positive: the surest way to success is understanding and avoiding the factors for unsuccess.

This book is both complementary to as well as a successor of my earlier book Insight in Innovation. It builds on the insights for success and adds the factors for unsuccess and practical guidelines for avoiding them. Both books are generic in nature, but whereas Insight in Innovation was primarily about innovation in large companies with a portfolio of innovation projects, this book relates to start-up companies with a single, significant innovative product under development. In a way the books are linked because Success in Innovation needs Insight in Innovation.

Part I

The Innovation Process

There is no standard recipe for success in innovation and each innovator has to create his/her own way that is fit for purpose, given the specific circumstances. But experience is an important factor for success in innovation. Part I describes a tested methodology for executing the innovation process that improves the chances of success.

1 Insights in Innovation

1.1 What is innovation

Innovation does not mean the same thing to everybody. Many people confuse innovation with invention and think that both words stand for the same thing. Others equate innovation simply with creativity or entrepreneurship or 'doing things better'. All these concepts are connected to innovation but do not represent the essence of innovation. Innovation is the process that develops an invention or a creative idea into a novel product and brings it to a user or customer, and that process is risky and needs an entrepreneur willing to take that risk. A practical, concise definition of innovation is:

'Innovation is bringing an insightful idea successfully to the market.'

Besides being short, this definition has several other attractive features. It brings across that:

- Innovation is not a one-off act or a flash of insight, but a process.
- Innovation needs insight. A creative idea by itself is not enough, but the new idea needs to be grounded in understanding its specific merits, such as the value to the customer and the requirements of the market place.
- It is not the technology but the market that determines the fate of the innovation. Market means here a collection of customers that use the innovative product, whether they have paid for it or not. This implies that customer needs and satisfaction must have a central position in an innovation effort.
- The word 'successfully' reflects a key insight: if the novel product is not adopted by (a sufficient number of) customers, it is not an innovation!

In its simplest format, a starting innovator has to achieve three things:

1. Develop a new product.[1]
2. Convince investors to finance the innovation effort.
3. Find customers that are willing to use (and pay for) the novel product.

This may look simple, but each step is a complex process in its own right and furthermore has to be executed at the right time and in the correct sequence. Many starting innovators put all their efforts in the first step because that is where they are good at and it seems to be the

[1] Throughout the book, the word product will be used to describe the innovative item that is being developed for use by customers. The item can be a physical product, a service, or a 'solution' = combination of a product and a service.

Success in Innovation

most important part of the innovation. As a result, many innovation efforts fail due to lack of money or customers and not because the product did not work. Finding customers that are willing to take the risk of buying and using an unknown product is the deciding factor for the success of an innovation. Considering potential investors as a first customer is not a bad start because also an investor has to believe that the product is worth spending money on.

Besides dealing with the above complex set of processes, the innovator has to also start and built a business that is capable of creating value with the innovative idea and extracting part of that value for the new venture. This means that, besides the three tasks above, the innovator also needs to carry out another task successfully:

4. Developing and organizing the capabilities for making and selling the product profitably.

The learning points are that developing the product is not sufficient, but simultaneously a vehicle, usually a business, for bringing the product to the customer has to be developed as well. Thus, an innovator has to be more than a smart scientist or creative designer but, in the first place, needs to be an entrepreneur, capable of developing and running a business. For a start-up company, the definition for innovation can be adapted to:

'Innovation is creating a successful business from an insightful idea.'

With this definition of innovation in mind, it is important for a techno-starter to appreciate how a true opportunity from the many ideas that present themselves can be recognized. Key characteristics of a sound innovation opportunity are[2]:

1. Creating value for customers by solving a problem or meeting a latent need.
2. Offering attractive profit potential.
3. Fitting with the capabilities of founders.
4. Being amenable to financing.

These characteristics may look obvious, but they provide a simple and effective screening tool for looking at innovation opportunities. In a way, the fourth characteristic is somewhat superfluous. Typically when the first three characteristics are satisfied, financing will not create an insurmountable hurdle. As the saying goes, 'there is always money for a good idea'. But it provides a good check whether the assessments of the first three characteristics were fair and realistic and not based on optimism and idealism. When this assessment is positive, the innovation project can start.

For understanding the intricacies of the innovation process, it may be useful to compare innovation to a journey of discovery or a quest. A quest shares many aspects with innovation and they can serve as learning points:

1. A quest has magical appeal, not recognized by most, but irresistible for some.
2. A quest leads through unknown and uncharted territory.
3. Most quests fail and never reach their destiny.
4. The 'treasure', if found, is usually different than expected.

[2] *Entrepreneur's Toolkit: Tools and Techniques to Launch and Grow Your New Business (Harvard Business Essentials)*, Harvard Business School Press (Compiler), 2005.

All these features can be found in the innovation process, including the often seemingly irrational drive that an innovator has, based on the conviction that the innovation is valuable and will be successful.

1.2 The Bridge Model

In abstract terms, the overall effect of an innovation quest can be described as transforming intellectual capital, embedded in formalized know-how and patents, and human capital, skills, and capabilities embedded in people, into a product that can create financial capital. Innovation is a process for creating and transforming value.

The bridge model reflects that innovation builds a bridge between the 'intellectual world of ideas, technology, inventions, knowledge, patents' and the 'commercial world of customers, markets, utility, value'. Figure 1.1 illustrates the concept of innovation as crossing a bridge between two different spheres: the intellectual world of science and arts, and the commercial world where innovation is measured in terms of 'utility' or 'value'. Most of the time that value will be expressed in money, but that may not always be applicable. In a commercial environment, creating value means generating attractive revenues. Not all innovations will generate revenue. Innovations that are only used internally, such as novel business processes or computer programs, may only be used for reducing costs. Innovations in or by nonprofit organizations may also not measure success in terms of revenues. The key requirement for success in innovation is that it is used by satisfied customers.

Planning and leading an innovation effort can be compared to planning and leading an expedition into unknown territory. The actual actions to be taken will vary starkly for different destinations, but they all have to go through the same process. An expedition to the South Pole, the top of a mountain, or the depth of a jungle may seem to have little in common, but in generic terms, they all go through the same phases: planning, preparing, the quest itself, and the way back. The fourth phase is an essential part of the expedition and should not be overlooked because reaching the top of the

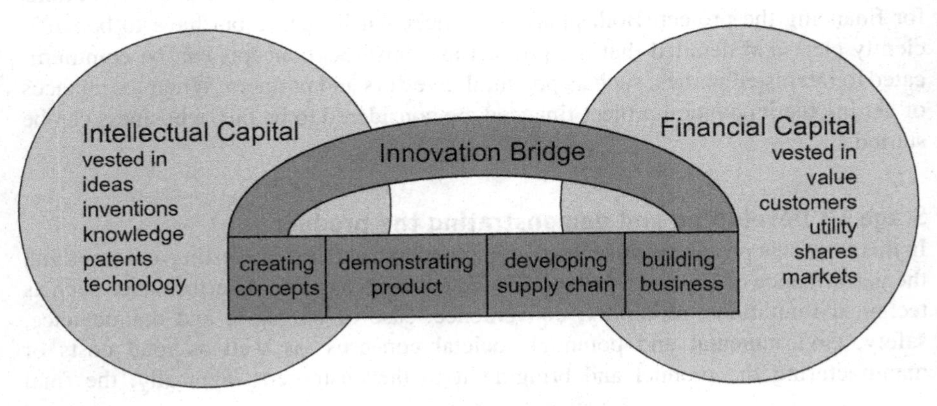

Figure 1.1 The innovation bridge.

mountain but not returning home safely does not constitute a success. Similarly, an innovation quest has four phases:

1. Creating the concepts for the product and the business.
2. Developing and demonstrating the product.
3. Developing the supply chain to the market.
4. Building a business with a robust cash flow.

Staging of the innovation process is important because in each stage the activities are different in nature and requires different capabilities. As a consequence, the people involved, including the leadership, may need to be changed when the innovation moves into a new stage. Understanding what has to be done in each stage as well as the how and the why, and doing them in the right sequence are key to success. The required changes with the shifts to the next stage have to be managed properly. The fourth stage is the 'proof of the pudding' and determines whether the innovation is successful or not. But not finding 'the pot of gold at the end of the rainbow' does not necessarily imply failure. If the quest is stopped in time and for the right reasons, the venture should not be considered a failure, but an heroic attempt that was worth doing.

Stage I: Creating the product and business concepts

In this stage, the innovative idea has to be developed into a business concept that includes a product development and a business plan. The product development plan describes the perceived advantages of the product in terms of benefits for the target customers, together with an analysis of the required capabilities, the timeline, and a cost estimate for the development. The product development plan is preceded by a conceptual design of the product. The design can be limited to the exterior shape of the product, but that is usually not sufficient. For radical innovation, the design is in the first place the interaction of the various elements and technologies, and the way the desired features are to be realized. Where possible, a functional model of the new product is an added bonus.

The business plan describes the target customers and potential revenue streams and includes a discussion on alternative business models and the possible options for financing the project. Both plans are conceptual in nature but have to be sufficiently clear and detailed that the product and business concepts can be communicated to interested parties, such as potential investors and partners. When the chances of getting the innovation project financed are considered to be fair, a business can be started.

Stage II: Developing and demonstrating the product

In this stage, the product is developed, demonstrated, and tested. Testing and assessing the performance of the novel product should cover all aspects of performance such as technical limitations, reliability, convenience, ease of operation and maintenance, safety, environmental and potential societal concerns, as well as total costs of manufacturing the product and bringing it to the customers. Typically, the final product at the end of stage II will differ from the concept product as envisaged in stage I and accordingly the business plan will have to be adjusted. The big question at the end of stage II is whether the product will be able to meet the needs and expectations of the target customers.

Stage III: Developing, or acquiring access to, the supply chain

In this stage, the path to the market has to be prepared in detail and executed. The core activity is to build up the capacities that are required for launching the new product into the market, and this will include aspects such as manufacturing the product, distribution and logistics, sales, and after sales services This does not mean that all these capabilities have to be in-house; it is usually better to follow a low-capital approach and create access to these capabilities via outsourcing or partnering. For example, the product can be made via toll manufacturing or distribution can be outsourced. The final, detailed business plan has to be created and agreed before the launch. The complexity of this stage is often underestimated by the innovator at the planning stage and the potential problems tend to be recognized too late. However, developing an effective and efficient supply chain to the market is critical for success.

Stage IV: Building the business

This stage is the final test for success because the product is now launched into the market and will be tested under real-life conditions and by real customers. Even when the product has been rigorously tested in pilots and test panels, this stage usually still provides one or more surprises and the business has to be ready to respond to feedback from the early customers quickly and efficiently. This requires building an organization that has the capacity, capabilities, and business processes to provide the customers the required services at the appropriate levels of cost, quality, and timeliness. This is the phase when the company has to start generating sufficient revenue for survival and continuity. If that has been achieved, the innovation can be considered as successfully completed.

The Innovation Funnel

Another model widely used in the innovation literature for describing the innovation process is the innovation funnel. The funnel has three stages and its shape reflects the concept that the number of innovation projects reduces when they pass through the stages in the funnel, because projects fail or leave the funnel (Figure 1.2).

The funnel model is useful when applied to managing a portfolio of innovation projects, as may occur in large companies, because it can be used for mapping the position and progress of the projects. However, in the case of a single project, which is the typical condition for a start-up company, the funnel concept is less appropriate.

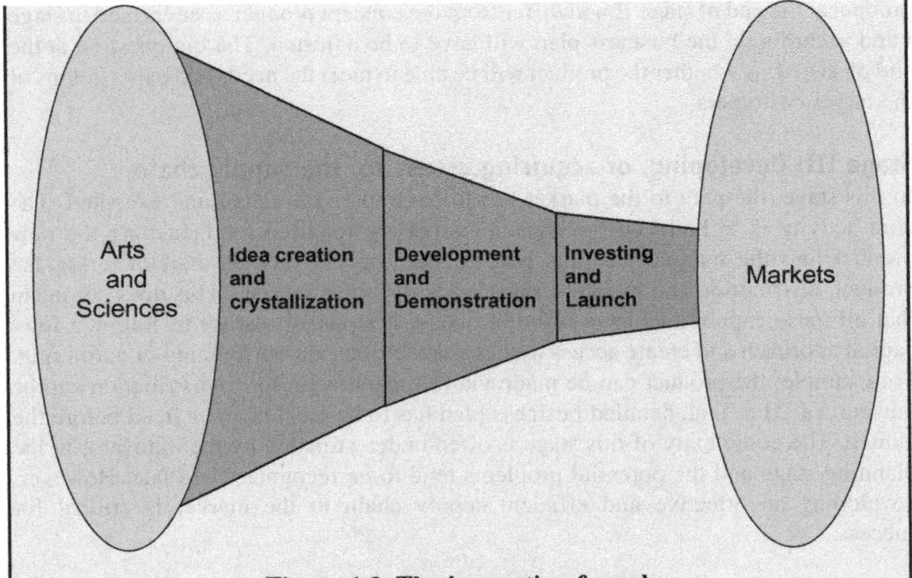

Figure 1.2 The innovation funnel.

The bridge model uses four stages and the practical advantage of a fourth stage in the bridge model is that it alerts the entrepreneur that the battle has not yet been won when the new product is introduced into the market. The measure of success is that the new product creates an adequate and robust revenue stream. A successful launch is not good enough, because many, if not most, new products disappear from the market within a year, even incremental innovations such as new varieties of established retail products.

1.3 Chances of success

Most innovations fail and that result is intrinsic to innovation. The typical success ratio for innovation is one success out of ten attempts, and this poor, rather unattractive, result is not only realistic, it is also not that easy to improve as a simple example will show. An optimistic techno-starter may believe that his invention has at least a 50% chance that it can be developed into a technically sound product. For a first-time innovator with a good idea, 50% chance of success may appear low, but for a theoretical idea without further evidence, 50% is a high estimate.

But there are two other hurdles between idea and success: the innovation process needs to be financed and the product has to be commercially successful in the market. Even if these two steps also have a 50% chance of success, the overall result is only one in eight. Furthermore, assuming that Murphy's Law also holds for innovation and something unexpected will go wrong with a chance of one out of ten, the chance of success of an innovative idea is $0.5 \times 0.5 \times 0.5 \times 0.9 \sim 0.1$. This result represents the numerical illustration of a basic rule of thumb in innovation:

Rule 1 – Take 1:10 as the chance of success in radical innovation

Of course, the figure of 50% chance of success for all three hurdles is unusual. If the invention has been tested at laboratory scale, the chances of technical success could be higher. But the chances that the whole project can be financed successfully or that an adequate number of customers will buy the product can be significantly lower. Embarking on a single innovation is a high-risk enterprise, and managing risk is one of the essential tasks in innovation. Large companies and venture capital providers can manage that risk by developing a portfolio of innovations that should be sufficiently large and diverse to have a number of winners. For a starter, this option is not available. At the beginning of an innovation project and without further specific knowledge, assuming a 50% chance of success for each hurdle is optimistic but is often done implicitly or explicitly. The challenge is to improve the chances of success and to know when to stop in time to limit the losses. For a starting innovator, managing for success means trying to improve the odds from 1:10 to, say, 1:4. This is not simple; it means that for each of the three hurdles the chances of success have to be improved from 50% to a level of about 60-70%. This leads together with Murphy's Law to the desired result: $0.7 \times 0.6 \times 0.7 \times 0.9 \sim 0.25$.

A probability of 70% for passing a hurdle successfully may seem quite achievable, but for complex, radical innovations, a 70% chance of success is high. For instance, the chance that a complex product or process will work in practice as planned, after it has been successfully tested in the laboratory but not yet demonstrated at larger scale as a test model or in a pilot unit, is only around 70%. Improving the chances of success in innovation means that each of the three hurdles in the innovation process has to be tested and secured at an early stage. It obviously helps significantly if the innovation effort can be financed without external investors or the strategic partner also is a launching customer.

The numerical examples also illustrate why big companies can adopt buying successful start-ups as their preferred innovation strategy. With the risks of the innovation much reduced or eliminated, the purchase price can be much higher than the development cost, potentially creating a win-win situation for both parties. Effectively the big company outsources its radical innovation effort to start-ups that are willing to accept the high risks, and it can focus its own innovation effort on incremental innovation for its core activities.

Considering and appreciating the chances of success, an innovator should ask him/herself two fundamental questions before starting his quest:

1. Is the idea good enough and worth all the effort?
2. Do I have it in me to be a successful entrepreneur?

The first question should have been addressed in financial terms in the concept business plan, but there are more considerations at stake for an innovator: the investment in time and adopting a 24/7 job without much reward for potentially an extended period. For the second question holds that having a good track record as a technologist or designer is useful but not sufficient. The fundamental advice for a starting innovator is given in rule 2:

Rule 2 – Think twice before you start

An honest assessment of the two questions can be very useful for avoiding unsuccess. If the answer to the first question is negative, the quest should not be started. A negative answer to the second question may be a good reason to find a trusted partner. Starting with a partner may well be the preferred approach because the time of the lonely innovator that creates a hugely successful business all by himself/herself is probably a thing of the past. Even Edison did not start by himself but created a team around him with engineering and commercial skills. The chance that one person has all the skills required for success is so slim that innovation rule 3 reads:

Rule 3 – Do not start alone

It is very useful to have a partner with complementary interests and capabilities. If you are the creative, nerdy engineer who can think with your hands as well with your head, look for a partner that has the commercial touch and business experience; if the partner happens to be a millionaire as well willing to finance the venture, then count your blessings. The latter condition is rare and starting should be a lean exercise.

And remember, if you have not asked and answered the above two questions yourself, the investor will ask the same questions and expect a clear and convincing answer. In abstract terms, an innovation needs to have a critical mass of intellectual and human capital for convincing an investor and being successful. Fortunately, a few of the required qualities are under control of the innovator: passion and perseverance; they are the key ingredients for success as expressed in rule 4:

Rule 4 – Passion and perseverance are the engines for success

Or in the words of Edison, *"Innovation*[3] *is 1% inspiration and 99% perspiration."*

Edison is well known as the inventor of the electric light bulb, but he should be recognized as the successful innovator of public lighting. He intuitively adopted all the right practices in his quest for bringing his insightful idea to the market and created a successful business out of his invention. His performance as an innovator is impeccable; he was both an inventor and an entrepreneur as the story below illustrates.

A perfect innovation

In 1802, Humphrey Davy discovered that an electric current passing through a metal wire could give light. In 1878, Edison joined the ranks of the many researchers trying to develop a light bulb with an acceptable operating life.

[3] The original quote reads "Genius" rather than "Innovation," but history has changed the quote.

In less than one year, he improved the lifetime of the bulb from minutes to 40 hours by using inert atmosphere inside the bulb. He obtained a patent and started the Edison Electric Light Company. Edison already had in mind that, if only electricity were to be distributed in a way similar to gas, electric light could become economically attractive. He improved the lifetime of the bulb to 300 hours and at the same time developed a dynamo that generated electricity at 110 V in combination with a small distribution network and demonstrated the whole system in 1880 with 425 light bulbs at his demonstration site at Menlo Park.

Edison Electric Light Company financed another demonstration project, but the other shareholders did not believe that electricity would replace gas for illumination purposes and refused to make the investments in the factories to make light bulbs as well as generation and distribution equipment. Edison decided to finance the investments himself with his shares in Edison Electric. He became an entrepreneur and created three new companies to manufacture the equipment. In 1892, Edison Electric changed its mind; the Edison Electric Illuminating Company was established and the new company obtained permission to light up Wall Street in New York. The rest is history.

This story illustrates the typical problems an innovator has to resolve:

- Developing, demonstrating, and commercializing a novel product.
- Financing and starting a business.
- Overcoming resistance to change.

In only four years, Edison brought a significant invention, based on a breakthrough technology and requiring a new infrastructure, successfully to market. Such a short time may well be a historic low for a disruptive innovation. Simultaneously, he also introduced a few new standards: 110 V for electricity distribution in the United States and the screw size of the bulbs.

Edison had a unique way of dividing a big problem into a series of small problems and solving them one by one, while keeping the whole value chain in mind. He approached the problem both from a technological and a business perspective, and overcame the resistance against a change from the existing practice, the gas lantern as street light.

The unique combination of qualities in one person — creative and analytical powers, sound entrepreneurship, and the drive to succeed — is the prime factor for this remarkable achievement. The above 'points of excellence' are archetypal and still very valid for today's innovator.

The achievement of Edison is more remarkable in light of the competition. Five years before Edison introduced electric street light to New York, the streets of Paris were already lit with electric arch lamps that were much brighter and with whiter light. However, practical and emotive customer preferences made the difference. Customers preferred the smaller, yellower light from the light bulb because it resembled more the traditional light of the candle.[4] Furthermore, Edison had developed a comprehensive, standardized infrastructure for his system, with switches, plugs, sockets cables, etc., much to the convenience of the customer.

1.4 The impact of innovation

Innovation can be seen as the third step in a process that starts with a scientific discovery, followed by technological invention, commercial innovation, and diffusion (see Figure 1.3). Diffusion is the step that follows innovation and relates to the process and speed in which the new product is adopted by society at large. When an innovation diffuses into society, it will affect the way of life, the more so the wider it has been adopted.

The idea behind this thought model is that new insights and ideas are cascaded to the subsequent stage and have to be processed in the technology and business domains before they can create value in society. This model seriously simplifies reality, but it serves to show that the impact and value of innovation have to be measured by its use in society and not by the efforts in the science and technology domains.

Most innovators will like to believe that their innovation will improve the quality of life and change the world for the better. Therefore, it would be fair to expect that their efforts will not only be rewarded in financial terms but also with warm appreciation. Indeed, innovators are admired for their efforts, but they should be aware that not everybody will be happy because there rarely are winners without losers. Innovation also creates losers and losers tend to be unhappy. The purpose of innovation is creating positive, valuable change with a new, better product that improves the quality of life for a customer. But, unfortunately, innovation always also has a negative element because other existing products and businesses will become less attractive. Sometimes this change is unintentional and can be seen as the collateral damage of innovation.

The threat of negative impact and of undesired change will create resistance to an innovation, and the more radical the innovation is, the more intense the resistance can be.

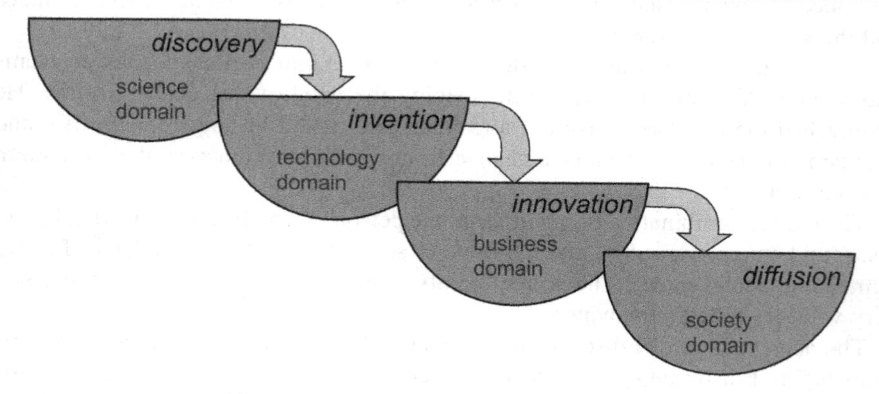

Figure 1.3 The impact of innovation.

[4] P. van Dooren, *En toch werkt het — uitvindingen waar je van had moeten horen*, Davidsfonds Uitgeverij, 2012.

Joseph Schumpeter, who is considered by many as the founder of the theory on innovation, argues that innovation leads to periods of 'creative destruction', as innovations cause existing technologies, systems, and equipment to become obsolete.[5] Well-known examples of periods of change from history are the transition from sailing boats to steam ships, or the introduction of cars and trains. Resistance did come not only from competitors using the traditional technology but also from sections of society that resisted the potential change in lifestyle.

Resistance to change is not limited to major innovations that have the potential to change the lifestyle; it can also happen to the small 'improvements of life'. The umbrella, invented by Jonas Hanway in the eighteenth century, was resisted on religious grounds because it interfered with the intentions of the Heavens. The current shift away from the traditional cork for sealing a bottle of wine is not welcomed by all wine lovers even when the new products seal better and are easier to use. The same people will be horrified by the invention that aims to change the taste of wine in a particular way by putting the bottle in a microwave oven for breaking specific nanocapsules that have been added. Attachment to tradition and nostalgic feelings are not the natural supporters of radical innovation.

Change should be seen as an integral part of innovation, and managing the resistance to change is an integral part of managing an innovation process. In change management, the term 'crossing' is often used to illustrate the steps that have to be made and the emotive reaction that a change in organization or working practice can bring. Both innovation and change involve crossing a divide, and both are considered risky activities. Innovation, change, and resistance are linked intrinsically: innovation creates change and change creates resistance.

Societal resistance is very dependent on the timing of the innovation. In certain periods, specific types of innovation are welcome and appreciated. For instance, at the beginning of the Industrial Revolution, mechanical innovations were popular; in the first part of the previous century, the development of chemical processes based on hydrocarbons from coal or oil was seen as progress; and in the current emerging digital era, novel ITC applications are welcomed by society at large for shaping new ways of living.

Being in tune with the prevailing value system of society is a major advantage and a key success factor. Technologies develop, mature, fade away, and are replaced by more advanced ones; they can become fashionable or lose their charms. The resistance against nuclear energy is based on a divergence with prevailing values in society; its perceived long-term risks and its association with a destructive, military technology from the past are more dominant than the vision of a constructive technology for the future. These days, new technologies have to be green and sustainable, and a growing part of society sees solar energies as a better technological option for the future.

But an innovator has to be in tune with more developments. Also investors go through cycles of interests and preferences. Asking for money at the wrong time may give an unnecessary negative result. Close contact with venture capital providers

[5] J.A. Schumpeter, *Capitalism, Socialism and Democracy*, 1942.

and knowing when to ask for money can make the difference between success and unsuccess. All these factors lead to a basic rule in innovation:

Rule 5 – Timing is of the essence

Success in innovation depends on recognizing and taking advantage of the 'window of opportunity', with respect to both customers and investors.

However, the other side of the coin of being in tune with societal and technological developments brings the risk that somebody else has a similar idea and is developing a competitive option. Brilliant minds work alike. And the winner may not be the best invention or the first one to market. History teaches that predicting which innovation will prevail is very difficult. Not only technical, commercial, or rational factors play a role, also emotional and irrational considerations are important. The competition between the options for powering a car at the beginning of the previous century may serve as an example. Steam power provided familiar and reliable technology, the electrically powered car was fastest and easy to operate and cars with internal combustion engines were noisy, polluting, and difficult to start, but the latter came out as the ultimate winner. This example illustrates another rule in innovation:

Rule 6 – The winner is unpredictable

Finally, it is important to appreciate the impact that radical innovation has on the bottom line. Contrary to popular belief, radical innovation is not the best way to get rich fast. On the contrary, the benefits of innovation tend to come slowly, and the more radical the innovation is the longer it will take. Radical innovation is a long-term investment for delivering a contribution to the future cash flow of a business, and incremental innovation is the vehicle for creating cash by improving margins in the shorter term.

The lead times of certain type of innovations can be very long. New processes in the chemical industry or medicines from the pharmaceutical industry make take 30 years before they make a genuine impact and contribution to the bottom line. New engines may take a decade or more. On the other hand, popular ITC applications can bring in real money within months. But here the life expectancy of the innovation runs the risk of being relatively short. In the music industry, vinyl LPs were replaced by tapes and later by CDs, which at the time were believed by many to be the final solution, but who buys a CD now? At the end of the twentieth century, the fax was seen as the ultimate communication system in the business world. The benefits of radical innovation to the bottom line tend to be 'late, but long lasting' or 'soon, but short', depending on the lifecycle of the technology or product.

1.5 Innovation needs to be managed

The message that innovation needs to be managed may well be the hardest message to get across. After all, innovation needs creativity, passion, and determination, and management with its business processes and bureaucracy can only stifle the effort and interfere with the dream. These concerns are valid and a management style that is not fit for purpose can easily do harm rather than support the innovation quest. Innovation needs its own type of management. The following quote is ten years old, but still very true.

> *"Innovation is a complex process and typically the road from idea to market is tortuous and full of unexpected surprises. For hundreds of years the innovation process had a large degree of randomness and this was thought to be part of the natural flow of the innovation process. But innovation can and needs to be managed. Innovation is not a magic ingredient that some companies have and others don't. It is a business process like credit management, strategy development or competitor intelligence that can be neglected or managed to its potential. However, probably because of its complexity, innovation is one of the last business processes still not subject to systematic management everywhere. Since the industrial revolution, manufacturing processes, investment decisions, project implementation etc. have been the subject of many management theories and have well-established best practices, but the history of innovation management only covers about twenty years. Drucker called the management of business processes one of the great innovations of the twentieth century, but innovation is one of the last areas where it is applied."*

The above quote from *Insight in Innovation*[6] is given to emphasize the importance of good management practices in innovation. But it should also be recognized that there is not one single methodology for managing innovation. Fit for purpose means horses for courses. In a large company, the main aim of innovation management could be to 'protect' the innovation team from the standard rules and bureaucracy applicable for the core business, and give them more freedom to do their own thing and follow the leads from intuition and exploratory experiments.

To a techno-starter, recently graduated, working in the cellar of a university supported by two part-time students, the need for managing his innovation efforts may seem somewhat overdone. It may look more attractive and efficient to follow an intuitive approach and adopt the 'trial-and-error' method. The trial-and-error method is a proven approach in innovation and, with sufficient passion, money, and determination, will often lead to success. The intuition of a brilliant developer is as important as systematic analyses.

However, an innovation process is a complex exercise, and the more complex the innovation is, the more that it needs to be managed for improving the chances of success. In most innovations, money is in short supply and controlling cost by minimizing the number of failed trials is an important factor. Project management does not represent the fun part of innovation, but it is an essential element. It is essential because the risks are high and the chance of success is low for an individual innovation project. The minimum result from understanding the innovation process is a shift from a reactive to a proactive mode of managing.

[6] J. Verloop, *Insight in Innovation — Managing Innovation by Understanding the Laws of Innovation*, Elsevier, 2004.

In the early parts of the innovation process, management should not be equated to organizing management meetings and introducing a slow and costly bureaucracy. It will be in the first place leadership in creating the vision and designing the way forward. It will also include transferring the passion and commitment to the rest of the team, knowing what to do at each stage and carefully planning the next step. Some innovators will do this quickly and intuitively correct, but for most, a structured process will be helpful and a safeguard against missing opportunities or taking on too high a risk.

A quote from Basil Springer makes the same point in a more expressive way:[7]

> *"In my interactive sessions with start-up entrepreneurs, as is now my wont, I introduce my new business management matrix tool as part of my triaging process (to borrow from a medical analogy) where I determine the priority advice to the entrepreneur based on the severity of the gaps in the 25 cells of the business management matrix.*
>
> *It is amazing the level of ignorance which is displayed by budding entrepreneurs who have good ideas but no knowledge of business management and no access to money."*

[7] Dr Basil Springer GCM is Change-Engine Consultant, Caribbean Business Enterprise Trust Inc. This quote appeared in the Barbados Advocate's *Business Monday* of June 18, 2012.

2 The Innovation Bridge

2.1 Crossing the bridge

The bridge model is not just an intellectual construct as part of a general theory on innovation, but it has also been designed to assist the innovator in understanding and managing the innovation process. The stages in the model give structure to the innovation process and make it more transparent. The model makes managing the innovation process simpler and more effective because each stage defines the things that have to be done and identifies the major risks and hurdles that could block or hamper the success of the project. Together the four stages form the necessary steps for creating a successful business from an innovation. The stages need to be kept separated, but there will be a degree of overlap because at the end of the stage preparations for the next stage will start, see Figure 2.1.

But not only the stages need to be managed, but also the transitions between the stages require focused managerial attention, because the organization will have to go through a transition phase, and a change process is emotive and risky. The tollgates between the stages serve as key attention points for assessment and decision making on whether to stop, change, or continue with the innovation project. The tollgates as reflection and assessment points are an integral part of the model. They are control points for safeguarding the correct progress of the innovation project with a set of rules designed to enhance the chances for a successful crossing.

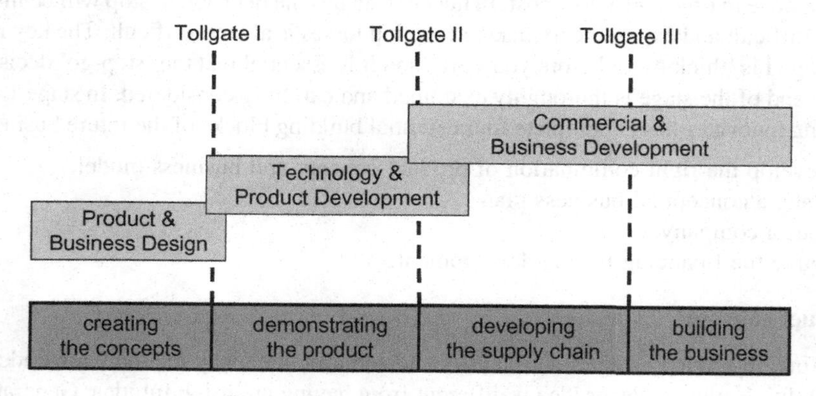

Figure 2.1 Stages and Tollgates.

The essential rules are:

- Understand the product and the business before moving from stage I to stage II. Carefully evaluate the ways and means for finding the money and the customers, and assess the chances of success realistically.
- Do not enter the next stage before a stage has been completed successfully and the business case has been reassessed.
- Before starting a next stage, the implementation plan needs to be agreed and financed.
- Each stage needs a different set of expertise, and access to the required capabilities should be secured in good time. Consider access via partners or outsourcing.
- Although the balance will shift, it is essential that in each stage the development receives input from all perspectives, technological, commercial, and business. Performance, cost, and design have to be in balance.
- Typically, adjustments to the initial concept are necessary in stage II, but in stage III, the product needs to be frozen and changes to the product should be by exception only.
- Make sure that the product performs and can be delivered reliably before launching it and be ready to listen and react to feedback from customers.

The above rules represent a set of best practices. Adhering to them will not always be possible, but if this is the case, it should be done consciously and measures should be taken for managing the associated increased risks.

2.2 Stage I — Creating the product and business concepts

Stage I of the innovation bridge is easily rushed through in the excitement to start the 'real' work on the innovation project. The creative idea is there and the innovator-to-be is eager to get started; after all, time is money and the competition is not sleeping. But rushing and shortcuts are serious pitfalls in stage I; it is a stage that needs adequate thinking time and reflection. It is the phase where stopping the project can be done in time and at low cost. In the later stages, the decision to stop will be much more difficult and expensive to make; accepting losses is always difficult. The key rule for stage I is 'think twice before you start', and it is essential that the 'stop-go' decision at the end of the stage is thoroughly examined and carefully considered. In stage I, the starting innovator has to complete four essential building blocks of the future business:

- Develop the right combination of product concept and business model.
- Make a conceptual business plan.
- Start a company.
- Agree the financing for the development.

Product concept

The first objective in stage I is to understand and articulate why the innovative idea is insightful. Having a clever idea is different from having an insightful idea. Generating innovative ideas is not the most challenging part of innovation; there is no shortage of decent ideas in the world of innovation. An insightful idea answers the issue of the 'why' of the product. It makes one understand the specific features of the innovation that will create value because it has special appeal and addresses a customer need better than other products. The advantages of these features will need to make the offering to the customer sufficiently attractive for making the switch away from the current products.

Radical innovation has to deliver a step change in performance and/or costs for success. Jos Paques, successful innovator and entrepreneur in the bioprocess technology, uses the rule of thumb that a novel technology must have the potential for 70-80% improvement. If the savings are only 20%, chances are that the traditional technologies will achieve a similar improvement over the period required for commercializing the invention.[1] Seventy to eighty percent may look too high, but there is a good rationale to it. A novel product needs a 15-20% advantage in cost and in performance for success, and if the competition improves by 15-20% and 15-20% of the potential improvement is not achieved, one needs a 70-80% potential improvement in cost and/or performance at the start of the development process. These figures are based on process technology where performance improvement of 2-3% per annum and development times of more than 5 years are quite typical. For other technologies, the figures will be different, but the message remains that the step change of a radical innovation must be significant.

A new product should not only be assessed from a business perspective but should also be looked at in its societal context. It is a basic requirement that the product does not have negative side effects, but it should also provide benefits to society at large and fit into the prevailing societal value system. It does not make sense to develop a new product that could be labeled by customers as 'unsustainable'.

Business plan

Once the product concept has been completed, a business model has to be developed that fits the product concept and creates most value. This will also include addressing two associated questions: is that type of business an attractive business to operate in and how strong is the competition? Answering these questions honestly tends to be difficult. Most innovators are impressed by the quality of their idea and as a result tend to underestimate the competition. It is good to realize that the existing competition holds a very strong trump card: it has already established a relationship with the customers!

The purpose of making a business plan in stage I is to explore how the product could be developed, demonstrated, and brought to market, together with a guesstimate of the associated costs and potential revenue streams. Other important topics include the estimated chance of success and identification of potential investors and partners. All this should be brought together in a concise conceptual business plan that describes how the value of the innovation can be realized. The business plan needs to be quantified with realistic assumptions on costs, margins, and market share, but detail should be avoided because such data will be unreliable at this stage and may instill a false sense of accuracy and confidence.

It will be useful to have two variants of the business plan because there are two different audiences: the innovator(s) and the investor(s). For the innovator and his team, the business plan should clarify whether the idea is any good, and for the investor, the business plan should create compelling interest for financing the venture.

[1] De Ingenieur, 02.11.12 and www.paques.nl.

In stage I, the business plan should not be a major quantification effort, but in the first place a story that creates excitement and engages the audience with the figures serving to illustrate that the story is credible and realistic.

Eureka

It is often assumed that successful ideas come from Eureka moments when the innovator experiences a profound insight in a flash. However, this happens to be the exception rather than the rule. Most innovative ideas are the result of fairly long processes where the would-be entrepreneur considers several ideas and business opportunities. It is quite unusual that the innovation stems from the first idea that came to the mind of the innovator. Often enough it is not the brilliant idea that started the process of innovation, but the desire to start a business that stimulated the innovator to start looking for opportunities and promising ideas. It is open for debate which way is better and has more chance of success: to start with the insightful idea or with the business opportunity. The important criterion is that the product concept is linked to the right business concept.

The key results of the conceptual business plan should include:

- The preferred type of business and its potential value.
- A realistic assessment of the development path, both in time and effort.
- A realistic assessment on the chances of success.

Finding the answers to these questions is not straightforward and can be done in various ways. If the business domain is well known, a desk study in combination with brainstorming sessions and advice from experts or consultants can be adequate. But in unknown territory, this may be difficult, and when figures have to be replaced by assumptions and hypotheses, the value of the business plan becomes questionable. Testing the market one way or another is very desirable in such a situation. If the product and market allow a number of low cost tests, a trial-and-error approach can be a good option. A successful example of such an approach is given by Jennifer Fleiss, who in 2009, together with Jenn Hyman, founded the company Rent the Runway, designed to solve the common problem *"a closet full of clothes but nothing to wear."*

Ditch the business plan[2]

Jenn and I had a rule from the very beginning to never write a business plan. We found that starting a business is more about executing a series of iterative tests, each test eliminating some of the risks that entrepreneurs face when launching a

[2] http://www.inc.com/young-entrepreneur-council/best-advice-i-ever-got-jennifer-fleiss.html.

company. Instead of sitting around and putting together a business plan based on a hypothesis, we knew we needed to test our idea on the ground to see if we could actually promote "renting" as a new consumer behavior.

We had to get scrappy, and we first tested out our idea by going to Harvard with a trunk full of dresses that we let girls try on and rent. Next, we went to Yale and rented out the dresses but did not let women try them on. For the third trial, we sent out a PDF to students that said, *"Call us if you want to rent this dress."* So each time, we were getting closer and closer to what our actual concept was — an Internet dress-rental site — to prove that it was really going to work.

If possible, it is always the best option to do tests in the market to find out which business model works best, and more reliable and convincing than any desk study. When the business model is clear and not too complex and there is no need for investors, writing a business plan is less critical, but can still be useful for identifying potential issues.

Besides identifying how the business model creates the most value, the business plan should also assist the innovator to answer the two critical questions:

- **Do I really want to be in that business?**

Creating and designing a new product is one thing; manufacturing and selling a product is a different business. For instance, if the main strengths and interests of the innovator are creative lateral thinking and technology development, the innovator should either transfer managing the company to a more commercial person or the business plan should show that the preferred option is to limit the business to a technology development company or sell the business after the product has been developed. At the end of the day, the business plan has to be in line with the aspirations and capabilities of the founder.

The answer to the question should include the issue whether the would-be innovator likes to be in the specific line of business; different types of businesses require different affinities and capabilities. Experience in the business at hand is an important asset, just as experience with starting a company is.

- **Is it worth the effort?**

The answer to this question has two aspects: a financial and an emotional one. For small, compact projects, the answers are usually straightforward, but for long projects, they may be more complex. Presumably, the business plan shows attractive estimates for the revenue streams, but the costs for creating the business could be high and the time for development long. The emotive question can be translated into: could I use my time and efforts better? The financial questions can be best answered by calculating the Option Value (see Appendix) and compare it with the costs and risks involved.

In case there is more than one founder, the business plan is an essential instrument to ensure that the individual aspirations and objectives are aligned and have found their place in the business concept. If that is not the case, a discussion for improving the required alignment has to be started or agreement reached to diverge and go a different way.

It is best practice that the innovator/entrepreneur does not write the business plan without external assistance. The average starting innovator will not have the expertise to assess all aspects and external advice will in general improve the quality of the plan. External input is important for a number of reasons:

- External eyes look at the product and opportunity more realistically and can make the judgments more objectively.
- A partner in the discussion can propose alternative options.
- Last but not least, the external input can dampen possible overidealistic and over-optimistic elements in the plan.

A degree of idealism and optimism are valuable elements of a plan; no innovator will start on such a quest without them, but too much of them are unstable starting points for a business and can be blockers for investors. It is essential to spend adequate time on the development of the business plan. It is the basis for the most important decision in innovation: pursue or stop and such a decision should be considered carefully. The concept business plan in stage I is not meant to be definitive and is not cast in stone. Most products in stage IV will look quite different from the original design. The key requirement of the concept business plan is to show that there is a realistic, attractive opportunity.

The second purpose of the business plan is creating interest with potential investors. The concept plan should have convincing answers to the questions that investors will ask. But keep in mind that the plan or presentation has to be concise because providers of risk capital can get so many requests that they screen new ideas in minutes rather than hours. The key questions to be addressed are:

- Does the product address a genuine need of customers?
- Where is the value in the company and is the business offering worth the investment?
- What are the risks and how will they be managed?
- Is the quality of the team, specifically of the chief executive, up to standard?

Confidence by the investor in the quality of the entrepreneurship is essential for a positive response. If the business plan has been considered sound and accepted by the investor, the negotiations can start as to how the company will pay for the required financing and the role of the investor in the company. This can be a time-consuming and rather involved process dictated by the way the investor runs its business. Usually, the investors will require a significant part of the shares and occasionally even a majority share. Besides the share of the equity, the investor may like to have additional influence on the running of the company, and possibly even request a seat on the board. This may create a problem if there is more than one investor. It is usually better to limit the number of significant investors. Each investor is a partner in the business and it needs time and energy to keep the relationship happy.

Starting a company

Before approaching investors, the new company should be formally established with its legal base. Having a formal company will facilitate receiving money from investors because it makes clear what they are buying. The value of the company essentially is the combination of intellectual property and human capital. Patents provide evidence that the IP is unique and relevant and thus have the potential of creating value. One of the most important factors for investors is the power of the management team, and it is important for them that this human capital is linked to the company. The usual way for achieving this objective is by giving shares and/or options to the management team with certain restrictions for selling or vesting. These restrictions can be time based, for example, vesting after one or more years or milestone based, for example, achieving substantial funding or positive cash flow. In most cases, the founders and initial management team will be the same and the founders will have significant parts of the stock. This stock should also have selling restrictions because an early, unexpected departure of one of the founders could be detrimental for the company. For all companies, it is important to connect the human capital to the company, but for start-up companies, it is essential; thus, there should be formal structures in place to incentivize key staff for staying with the company. On the other hand, the ties should not be so strong that a desirable departure would not be possible because one thing is for sure: change will happen and the new conditions may require different solutions to those expected initially.

Most companies will also have a board for supporting the executive team. The board can also be useful for impressing potential investors. However, a not uncommon mistake made by techno-starters is to appoint only respected professors and scientists on the board. That may support the technological credibility of the company but can be rather counterproductive in the commercialization efforts of the company, because the focus in the company will not change in time. Start-ups could consider having an advisory council for scientific and technological issues. The quality of providers of advice and services can be a relevant consideration for investors and having a reputable board, accountant, legal counsel, and patent attorney, preferably known to the investor, can be a valuable asset.

Financing stage I

Typically investors will finance the innovation efforts in stages II and III. Financing the efforts in stage I usually comes down to various forms of bootstrapping. The efforts by the founder(s) in stage I are typically done for free and the external costs are covered with own money or from friends and family. A part-time job may be needed to cover the cost of living. Sometimes, seed money can be available from angel investors or as part of the service package of incubators. Crowdfunding can be an attractive option for innovations that have popular appeal, short time frames, and limited financial requirements.

Crowdfunding

Crowdfunding describes the collective effort of individuals who network and pool their resources, usually via the Internet or social media, to support efforts initiated by other people or organizations. Crowdfunding is a relatively recent method of financing and initially it was mainly used in the creative sector or for supporting benefit actions.

Crowdfunding can also be used by start-up companies by selling small amounts of equity to many investors. It is usually done for limited amounts of up to a few 100 k$ and for socially relevant projects. This is important because the investor has to feel emotionally attached to the purpose of the investment opportunity and for that reason is willing to accept the high risks, the limited influence, and the often low returns. Besides a financial reward, a company can also offer products or services in return.

Crowdfunding has additional advantages for a start-up because it provides a link to potential customers, creates a network, and gives an indication of the interest in the product under development. The disadvantage can be that maintaining the communication with such a large group of investors can be time consuming and distractive. Also the protection of IP can be difficult and the idea once published on the Internet can be easily copied.

Crowdfunding needs to be done via a proper platform that safeguards the interests of the investors. Investors will need to be supplied with a digital contract. Typically, a minimum amount is set as a threshold, and if that minimum is not reached, the investors will get their money back. Kickstarter is a well-known platform for crowdfunding and has funded about 30,000 projects with almost 3 million people since its launch in 2009.

2.3 Stage II — Developing and demonstrating the product

The second stage of the innovation bridge is the development and demonstration phase, which can take many different formats and may last from a few months to many years. During the tollgate I assessment, the direction and objectives for stage II have been set and agreed, and the task ahead is to transform the novel idea or invention into a commercially viable product. The development of the product has to be done simultaneously on two parallel tracks: technological and commercial. Characteristic activities in stage II are:

- Development and testing of the required technologies.
- Assessing the commercial feasibility of the product in all its aspects: production costs, logistics, distribution channels, safety, ecological impact, societal acceptability, etc.
- Demonstrate and test the product with a working prototype or pilot.

It is not unusual that the innovative concept will change a few times in stage II for a variety of reasons. For instance, the technology does not perform as expected or turns out to have undesirable features, the strategic partner shifts position, or the anticipated market changes. At each point that a significant change in the product scope is considered, the viability of the innovation should be reassessed. The case study of a super clean diesel fuel provides a real-life example of an idea that was grand at the start, but in the end turned out to be not viable.

Sootless Diesel

Diesel engines emit soot and these very fine particulates are a health issue. It is technically possible to create a sootless diesel by adding sufficient oxygen to the hydrocarbon fuel. A hydrocarbon with a high oxygen content such as methanol burns without emitting soot, and ethanol with a slightly lower oxygen content still burns almost without smoke and can therefore be used indoors in fondue burners and fire places.

The idea of a sootless diesel has a strong marketing appeal because it can eliminate or reduce the health concerns and the negative image of this fuel. Around the turn of the century, a global oil company started an innovation project for creating a sootless diesel at a competitive price. It was assumed that a slightly higher retail price would be acceptable because the product would have strong customer appeal, and the market for this product could be enormous with worldwide application.

Combustion experiments showed that the addition of small amounts of simple oxygen containing liquids such as ethanol or water did reduce soot emissions but not sufficiently for creating a sootless diesel. The amount of these liquids that can be blended into the diesel is limited because they do not mix very well. As a result, marketing suggested a change to the objective of the project to allow the use of more complex chemicals that could be blended without restrictions with the diesel fuel and focus on a different market in which a higher premium would be acceptable, such as heavy trucks and buses with the most serious emissions. With this approach, it was possible to create a sootless diesel, but the costs were deemed too high.

A shift of the target market to inner city traffic where the problems of soot were most restrictive and a higher premium could be acceptable appeared not to be attractive because this market was too small for the substantial development and investments costs. Furthermore, it became doubtful whether the overall environmental footprint of this fuel would be smaller than conventional diesel considering the energy and costs involved in manufacturing the complex chemicals.

With each step, the Option Value of the innovation reduced and it became negative for the final step. As a result, the research project leader proposed that the project be discontinued.

Such an outcome from an innovation project should not be labeled as a failure. The development of the product and the assessment of the value of the innovation had been carried out properly and the right decision was taken at the right time. Edison's words of wisdom express this condition accurately: *"I have not failed. I've just found 10,000 ways that won't work."*

As observed earlier, it is essential that the technical development activities in stage II are not ruled only by science and technology but are also steered by business and commercial considerations. This seems obvious, but it is often not done or done insufficiently because the relevant experts are not available inside the company and as a result the issues are not recognized. It is important that during the technical development all aspects that are important for the commercial success are thoroughly examined, not only the obvious ones such as manufacturing cost but also reliability, serviceability, and the overall economics of the total supply chain.

However, correct analysis may still lead to a difference of opinion on the right way forward. From a business perspective, innovation needs to be a linear process, but a researcher will prefer an iterative process with a recycle loop so that the project can be restarted from a better position and incorporate the learning points from the earlier approaches. Another source of conflict can be that the researcher would like to continue with improving the performance of the new product, but the entrepreneur prefers to freeze the development and is willing to adopt the 85% solution in order to keep the cost down and not lose too much time. In a small start-up company, the researcher and the entrepreneur may well be the same person and that person should be aware of these dual responsibilities and which one has priority at what time.

Stage II developments may need the involvement of external partners because one or more of the required capabilities will not be available in-house. For a start-up company, it will be unusual for all the engineering and commercial expertises to be available in-house. Creating cooperative arrangements with external parties either on the commercial or on the technology side is almost standard for radical innovation. The main advantage of cooperation is the reduction in the development costs, risks, and time. Developing all the required capabilities in-house will typically take longer and may not be successful. The main disadvantage of cooperation is the time and effort it takes to create a win-win platform and maintaining that platform until the task has been completed. It can take a long time before the parties understand each other's language, have built sufficient mutual trust, and established procedures to safeguard the use of intellectual property and avoid contamination.

A successful demonstration of the product at semicommercial scale leads to the end of stage II. Successful demonstration means that not only the research staff but all the relevant parties are happy with the outcome. The parties involved should include external parties such as partners and potential customers. For partners this will be obvious, but it is equally important to include customers. In fact it is a factor for success to include customers as early as possible in the development of the product, as expressed in rule 7:

Rule 7 – Involve the customer early

It is of great advantage if the demonstration is unambiguously successful and there is no further need to adjust the product. The latter is an exception, but it makes a big difference if the changes are only of a minor nature and do not change the key features of the product.

The volatility and uncertainty that are intrinsic to the development of a radical innovation project and the many hurdles that have to be taken can make stage II a treacherous domain. Not untypically, the project will need additional funding as a result of setbacks and changes in direction. The development effort in stage II tends to be underestimated, either out of ignorance or for making the business case look better and more attractive to investors. But the reality is that the development is always more difficult than expected and hampered by unexpected problems. Cautious planners will adopt the time-tested rule of thumb for radical innovation:

Rule 8 – It takes twice as long and it costs twice as much

Rule 8 may well be the most important guideline in developing a radical innovative idea.

2.4 Stage III — Developing access to the supply chain

The first thing the new team leader should do at the start of stage III is to sit back and rethink the project, and then review and update the business plan taking into account the results from the product demonstration and adding detail to the requirements for delivering the product to the market. The review will include identification of the new capabilities required for business development and commercialization, and the options for creating access to the supply chain. The key issue will be whether to invest in fixed assets for manufacturing, distributing, and selling the product or to develop partnerships for creating access to such facilities. If the choice is for the former, capital intensive option, the business plan will have to show that the required investments are economically sound and can be financed.

During the development process in stage II, the emphasis will have shifted from R&D to design and engineering. In stage III, the lead in the innovation quest will shift further and have to come from the business development and commercial side and this makes it important that the leader of the innovation project is changed. This does not mean that the role of research is finished; it has to remain involved but at a reduced level to be able to adjust the product when commercial considerations or customer feedback so dictate.

The next step is to build up the team with the new capabilities. For a start-up company, this will usually mean changing the capability mix of the whole company. This will lead to difficult decisions. Money is limited and bringing in new staff with manufacturing, logistics, and marketing expertise may have to be done at the expense of technical staff. The option to start another innovation project to keep the key technical staff happy and occupied is alluring, but this option could be in conflict with one of the key rules for a start-up company:

Rule 9 – Do not start a new project before the first one is a success

In general, success will mean generating a robust cash flow. Starting another project before generating adequate revenues will drain the working capital even faster and this can create a serious threat to the success of the first project and consequently of the whole enterprise. But even if the second project can be adequately financed, the new project can take away too much time and attention of the management team to the detriment of the first project.

One option for keeping the R&D staff could be to start developing Mark II of the product if there are good leads that the performance can be improved or the cost reduced significantly. But this option will also increase the cash burn of the company and should only be considered when there is adequate working capital. Another option is doing technological development work for other companies. This can be an attractive option that allows maintaining a high level of technical expertise without creating a liquidity problem, but care should be taken that it fits the business plan.

The above human resource and capability problem can also be approached from a different angle. Rather than build all the required capabilities to full strength in-house, it may be better to look for a partner that has the required capabilities and limit the in-house commercial staff to a core level. Finding a partner can have several advantages:

- The partner will already have access to the target market or other valuable assets in the supply chain, for instance, manufacturing or distribution facilities, and in this way reduce capital requirements.
- It will reduce the time to market, assuming the search and negotiation process does not take too long. And reducing time to market will increase the value of the innovation.
- A good partner can reduce the chance of making mistakes, assuming the partnership is desired by both parties and communication is and remains good.
- A good partner will in general be a strong point for convincing the provider of the required risk capital.
- One of the best things that a partner can bring, is being a launching customer.

A good partner will improve the chances of success considerably. Partnership can take a variety of shapes (see Section 7.4). The simplest option is a strategic cooperation whereby the partner provides access to or handles part of the supply chain. Another is creating a joint venture for the shared activities and the most extreme option is a complete takeover.

The key challenges for the commercial staff in stage III are assessing the market conditions and understanding the business and competition. The former can be done with the feedback from potential customers on the final prototype product or from a pilot test in a small market. Probably the best way for assessing the market is finding a launching customer that is willing to be the first user of the product, usually against the benefit of a discount or a degree of exclusivity. Not only can a lot be learned from the launching customer, but also it will facilitate and accelerate the introduction and penetration of the market. A launching customer is a prime strategic partner.

Other preparations will include establishing quality control systems, conditions of sale, business procedures, terms for doing business, and recruiting and training sales staff. The basic aim of all these activities is to be able to launch the new product into the market without a hitch. Obviously, being able to meet the demand at minimum cost is important, but avoiding mishaps and negative news is more important because that can easily make an unsuccess of a successful launch of a good product.

The challenge for the technical staff in stage III is to hold back on further product development. Although the demonstration tests with the product were successful and showed that the product was fit for purpose, undoubtedly points for improvement had been identified as well. In principle, the product as tested and approved should not be changed because that gives the best chance that the product will work as designed and promised to the customer. But comments from customers should be considered. A successful demonstration is not a guarantee that it will work in practice. Customers expect a high degree of reliability and ease of operation. Fifty or hundred years ago, customers were used to factory-produced goods such as cars, radios, and washing machines that were needing repairs regularly and that operating them needed experience. Customers today expect that operating the device should be simple and self-evident, as well as trouble free during the lifetime of the product.

2.5 Stage IV — Building the business

The proof of the pudding is in the eating. This time-tested saying still holds very true for innovation. Only the response from the market place decides the fate of even the most exciting innovation. Remember rule 6: The winner is unpredictable. The race with the competition has now started in earnest. The key objective after entering the market is creating a cash flow. As in all previous stages, liquidity is the determining factor for survival and success. The money that was raised in the previous stage is running out and returning to the shareholders or partners for more money is not an option; they are looking for dividends and profits, not for further dilution or more expenditure. Generating revenues will give access to a new source of working capital: the banks. With sufficient liquidity and the capability for paying interest, banks may now be interested in giving loans. Generating and growing revenues will not go without increasing costs, but that is allowed as long as the costs increase for that purpose only and not because of growing overheads or generous bonuses. Keeping costs under control is essential (see Section 6.1).

Building a relationship with the customers is equally important. Feedback from the early adopters of the innovative product has to be incorporated for ensuring that the performance of the product and the service package supports the building up of a good reputation and attracts other categories of customers. Expanding the company with new customers, and maybe new markets, is needed to turn the initial losses into robust profits. Once the company has:

- a robust cash flow,
- a growing customer base,
- loyal and experienced staff, and
- a loan from the bank,

the innovation can be called a success. A loan from a bank is a good criterion for success because it shows that a critical third party has made the judgment that the outlook for robust revenues and margins is good. But, of course, not needing a loan is even better and not getting a loan is not necessarily a sign of unsuccess because banks can be 'unreasonably difficult' in times of crisis and austerity.

But although the innovation is now officially a success, there is no time to slow down because the company still has to show that it is there to stay. Statistically, the company is still in a danger zone in the early years; about one in three of new establishments fails in the first two years and more than half in the first four years.

2.6 Guarding the tollgates

The previous sections will have created the impression that innovation follows well-structured processes with the various stages clearly separated by well-defined decision points. In the real world, the innovation process, even if it is well managed, will be fairly chaotic. This is partly by design because preparations for the next stage have to start before the earlier one has been completed and partly because things do not work out as planned for internal or external reasons; for instance, development is taking longer, financing not forthcoming, and problems with hiring new staff. Such events cannot be avoided, but the value of the staging is that one is aware in which phase of the process the innovation is and which targets should be completed before new commitments are made.
The tollgates between the stages serve three purposes:

- Reduce the risk.
- Change the team.
- Reassess the business case.

There are several 'good' reasons to stop an innovation project:

- The product will not work or its performance is not good enough.
- The price of the product is not competitive and will never make money.
- The markets for the product are no longer attractive.

Any innovation project has the intrinsic risk that it has to be stopped: if stopped by default, it has failed; if stopped by design, it has not. Determination to continue against the odds has to be balanced with a willingness to stop. Table 2.1 lists a number of criteria for the stop-go decision at each tollgate. The list is not exhaustive, but it serves to illustrate the variety of tests that have to be passed for a comprehensive tollgate assessment.

Tollgate I is a good point to stop if the results indicate that way. Stopping at gates II and III is usually caused by failure of the product, technically or commercially, or by withdrawal of partners or investors.

Tollgates between the stages are the points in the innovation quest for assessing the chances of success and to make the stop-go decisions. But there are more critical points in an innovation project that need to be recognized. Besides the obvious ones such as formally starting the company and the time of the product launch, there are two points that are easily not managed properly: the point of product freeze and the transition from a technology to a commercially driven development company. Early in

Table 2.1 Tollgate criteria

Tollgate I	Tollgate II	Tollgate III
• Benefits for target customers identified • Concept business plan completed • Intellectual property position evaluated • Preliminary commercial and technological assessment • Project plan for stage II agreed and financed	• Product demonstrated with working prototype • Customer value proposition and target customers defined • Business plan prepared and adopted • Positive safety and ecology assessment • Access to all required capabilities • Project plan for stage III agreed and financed	• Response from customer tests are positive • Comprehensive and detailed business plan accepted • Access to total supply chain secured • Measures in place for securing adequate liquidity

stage III, the product has to be defined in its final format and changes are no more allowed unless dictated by external requirements. The transition from technology to commercially driven development creates a problematic period for the company with the changes in team composition and leadership, and managing this period requires both sensitivity for possible frustrated ambitions of some of the staff and a firm commitment to implement the required changes. Figure 2.2 illustrates the various changes and decision points that are part of an innovation project.

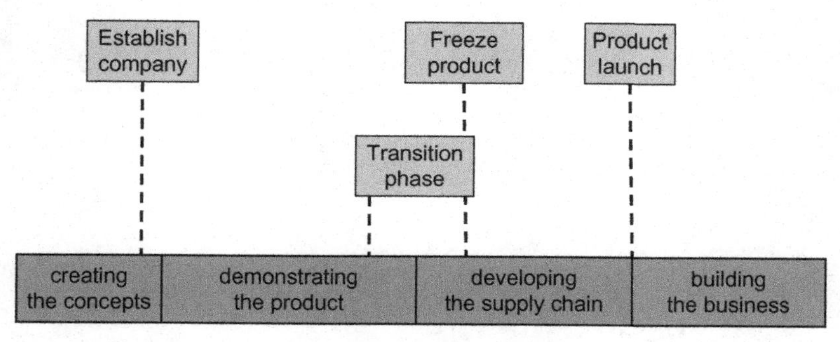

Figure 2.2 Critical points on the Innovation Bridge.

3 Building the Business

3.1 The building blocks

The problem that any techno-starter faces is that success in innovation involves building and running a successful business. Building a business requires a different set of skills from developing an innovative product and not every starter will master those skills. Building a business involves creating an organization with a staff, a structure, protocols and procedures for channeling the activities, keeping track of the money, and enabling management to steer.

There is a set of activities that can be considered to be the building blocks for growing the value of the company. They are related to understanding what to do, the competitive position of the company and possible exit opportunities, and developing realistic plans for executing, financing, and organizing the innovation quest. These plans are not one-off studies, but they are 'living documents' that need to be updated and reassessed regularly, at least in each stage and when the product definition has changed or the preferred business opportunity has faded away. Neither do they have to be extensive documents; they can be as basic as a checklist as long as the logic and actions are agreed and understood by all involved.
For a starting company, the key building blocks are:

1. Strategic orientation: identifying the preferred business opportunity as well as exploring the potential business and exit options.
2. Designing the business: defining the 'how', 'what', and 'who' of the business, including the revenue model and the key cost factors.
3. Project execution plan: agreeing the required resources for the technical and commercial challenges, and developing a timeline for the implementation.
4. Financial planning: understanding how much, how, and when capital has to be raised.
5. Organizing the business: creating a team and developing an effective and efficient set of procedures and protocols for executing the business objectives.

3.2 Strategic orientation

The driving force for strategic orientation is the innovation wisdom that the outcome will turn out to be different than the original idea, as expressed in rule 10:

Rule 10 – Even the best innovation concept is partly wrong

Appreciating that the outcome of the innovation quest is likely to be in one way or another different from the original idea is an important concept that should be ingrained

Success in Innovation

in the mind of an innovator. It should give the flexibility that is required to respond to unexpected changes during the project and also recognize alternative opportunities that may emerge. Success in radical innovation often depends on taking an opportunity when it is there. The challenge is to recognize the opportunity when it passes by.

At the beginning of the innovation quest, the number of opportunities that can be identified are basically limited to what to sell and how to sell. It is relevant to realize that rather than selling the innovative product, the business strategy can also opt for attempting to sell the IP sec or the company, that is the combination of intellectual and human capital. The advantages of these two options are that they are lower cost and lower risk, and can be an early opportunity before the product launch. These options are both business opportunities as well as viable exit strategies.

Selling IP in the early stages by itself is not easy even when patents have been granted. Patents and know-how without a working demonstration unit have limited value. Before the product has been tested, the value of the patents and know-how is more related to the cost of development rather than to its potential value as a business opportunity. Licensing or selling the know-how after completing stage II can be attractive when the next stages appear to be very expensive or difficult with the associated risks and problems for financing. The product can then possibly be better commercialized by an existing company that already has the capacity available for bringing the product to the market. In this way, a win-win situation can be created when the value of the innovation is higher for the existing company than for the start-up company and a deal can be made that is attractive for both parties. Usually, the buying party will determine the scope of the deal: buying the whole business or only the know-how and key staff. The room for negotiation for the start-up company will depend on its financial strengths, and in a weak position, the company may be forced to accept any deal that is better than going down.

Once the product has been launched successfully into the market, the options of licensing or selling the business become in general more attractive and valuable. Licensing to other companies can create additional value when these companies service markets or locations outside the reach of the start-up company.

If the base strategy has been selling the product, there still remains another important strategic choice: going for a B2B (Business to Business) or a B2C (Business to Customers) business. The margins in a B2C business may look very attractive in general, but a B2C business is more complex and difficult to establish than a B2B one and it also requires more skills and capital. In times of austerity, it may be hard to find investment money for B2C sales channels. The latter is less of an obstacle with the option to start a Web-based sales capacity.

Figure 3.1 presents the generic options that can be considered and assessed in the initial business plan. Usually, the base case will be taking the product to the customer, but it is important to be aware of alternative options because they can become more attractive than the base case or serve as exit strategies in case the base strategy can no longer be pursued.

A particular option may enter the process more or less unexpectedly and is difficult to plan for. A large company may make an offer to buy the start-up. Many large companies

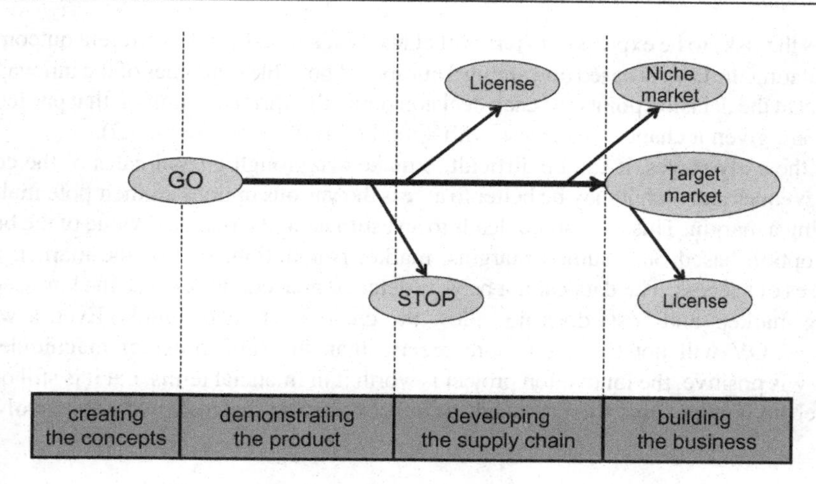

Figure 3.1 Generic business opportunities.

consider buying successful start-up companies as their preferred innovation strategy. The preferred timing tends to be once the technology has been demonstrated and the market tested, but the company can decide to do it earlier to be ahead of competition or for keeping the price down. Even when a premium has to be paid for purchasing the company, the cost may not be higher than for an internal development, because costs in a large company are structurally higher than for a start-up. This option is often attractive for both parties because the value of an innovation can be much higher for a large company than for a starting company and part of this additional value can be shared. A large company may be able to accelerate the growth of the product significantly because it has the required resources — financial, marketing, distribution, and sales points — with much better access to customers at lower cost with its existing facilities.

Start-ups may have this option high in their expectations, but it is difficult to base a successful business strategy on this option, because the key parameters are subject to outside control. The best strategy tends to be building a successful company that creates an interest from the big players.

A proper understanding of the strategic options is important for evaluating the value of the business. The usual method for calculating the value of an investment is the Net Present Value (NPV), but that approach is less appropriate for an innovation. The NPV is a useful tool for calculating the value of a well-defined project with a known risk profile that allows the inclusion of the risk element in the discount factor. The value of this factor should be based on experience with similar projects rather than being just an arbitrary addition to the cost of capital and inflation elements. Since an innovation project typically cannot benefit from the experience of similar projects, an NPV calculation is very arbitrary. The Option Value (OV; see the Appendix for more details) approach has two significant advantages.

Firstly, it allows for multiple outcomes of the innovation project such as early failure, a change in product and associated target market, or selling of the business. Secondly, it allows the risks to be expressed in terms of chances of success for the different outcomes. Calculating the OV[1] is based on valuing a number of possible outcomes of the innovation project at the decision points. At each decision point, the various options at that particular point are given a chance of success with a total of 100% (see Figure 3.2).

In the early stages, it can be difficult to make even rough guesstimates of the costs and revenues and then it may be better to assess the various options on their potential for creating a margin. This assessment leads to an estimate of the potential value of the business option based on assumed margins, market penetration, size of the market, and chances of success. The data cannot be very detailed or accurate because the knowledge of the markets and costs does not allow the creation of such figures. Even a well-assessed OV will not be much more precise than the right order of magnitude. If the OV is positive, the innovation project is worth it in financial terms, but it is still open for debate how positive the OV needs to be, considering the costs and risks involved.

3.3 Designing the business

In stage I, the conceptual business plan was made for convincing founders and investors that the innovative idea had the potential for becoming an attractive business opportunity. In stage II, the innovative product concept has to be developed and demonstrated, and simultaneously, a business has to be designed that can extract value out of the opportunity with acceptable and manageable risks. In the strategic orientation the various business opportunities have been identified and assessed, and the

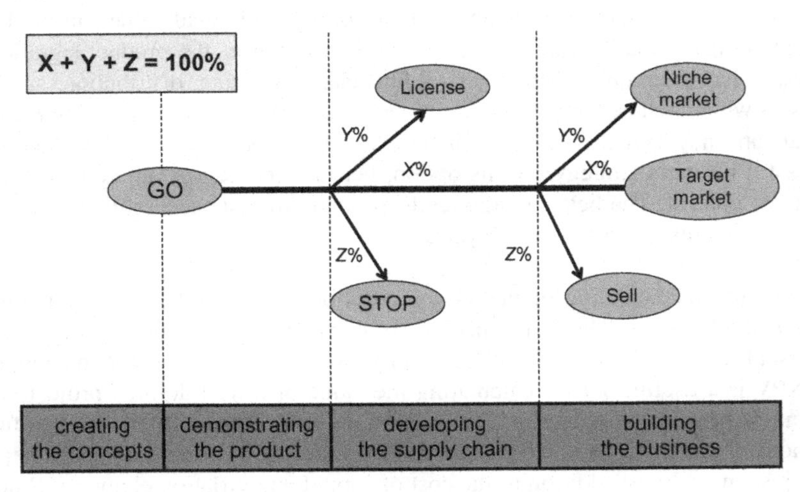

Figure 3.2 Modeling the Option Value.

[1] This method for calculating the OV is not rigorous and for that reason the terms Simplified Option Value or Risk Adjusted NPV are sometimes used. However, the term OV is preferred, because innovation is creating options for future cash flows and the suggested method assesses the value of that option.

preferred opportunity has been selected. The business plan has to design the business model and develop the business case for the preferred opportunity, as well as addressing the implementation issues and possible exit strategies.

A good business plan makes clear what needs to be achieved and what has to be done and can convince interested parties that it can be done successfully. Typical issues are:

- Which customer problem the innovative product is addressing, why, and how.
- What sort of business the company is going to create.
- The size of the market and the target customers.
- The required capabilities and finances, and how to get access to them.
- The technological, commercial, and execution risks and how to contain and manage them.

As was the case for the conceptual business plan in stage I, the detailed business plans in stages II and III will also have an internal and an external audience. The internal version is a communication tool of the management team to the staff of the company for aligning and providing focus to all the activities throughout the company. In stages II and III, the external target audiences are not only investors but also the potential partners that the company may need for successful implementation. It represents a major advantage if the development effort in stage II has been adequately financed and investors do not have to be approached for additional financing, because the conditions for refinancing at the end of stage II tend to be less attractive than in stage I (see Section 3.5 on Financial Planning and the Valley of Death). When the plan is presented externally, it is good to appreciate which issues are important to potential partners:

- Does the product look good and fit their own business?
- Are the financial figures realistic and attractive?
- Can the cooperation be mutually successful?

A comprehensive business plan does not have to be an extensive document. In principle, a slide presentation with a good story can be adequate and efficient. Obviously, the internal and external versions will be slightly different and emphasize different points.

One of the key issues to be addressed in the business plan is how to handle the access to the supply chain. During the strategic orientation, the preferred business option has been selected, but that had been done on tentative figures and emotional preferences. The business plan in stage II has to underpin this choice with a more detailed assessment based on actual performance of the product and better estimates of costs and competitive positioning. There are four options to select from:

- The company builds its own supply chain.

This option will in general have the potential for creating the most value for the company, but will also be expensive and difficult to implement.

- Access to the supply chain is achieved via strategic partners.

This option avoids the cost of investments and building the required capabilities by sharing part of the added value of the product. It can also reduce the risks and the time to market and thereby improve the chances of success.

- Activities in the supply chain are outsourced to strategic partners.

This option is similar to the previous one but reduces the own activities and requirement for in-house capabilities to a minimum. The disadvantage is the lack of influence on the effectiveness of the supply chain with the resulting reduced share in the added value.

- The business scope is limited to product development and monetized via selling or licensing the technology.

This business option of being a technology company avoids the need for a stage III effort, but it does not necessarily avoid the need for involvement in the supply chain, because the customers will want support for optimization of the supply chain. This is a low-capital option, and although the revenues can be modest compared to the other options, the profitability in terms of ROI (return on investment) can be very high. Longer term, this option can only be successful for either very unique technologies or if there is a continuous stream of innovative products.

Another issue is to determine which capabilities should be in-house and how they can be sourced and maintained up to the required standards. Key activities such as product development require in-house capabilities and typically the R&D effort is delivered by own staff. Even if a set of activities is outsourced, there is a need to have in-house expertise for selecting the best option and 'smart buying'. A potential danger is that keeping a small expertise unit up to standards can be difficult, and unique expertise can be lost to the company relatively easily.

The dominant issue for the business plan in stage III is how to launch the product. There are three generic options ranked in order of decreasing risk:

- The product is launched and sold to whoever wants to buy. The risk associated with this option is that if the product is not yet perfect from the start, the negative news will spread and the reputation of the product is damaged beyond repair. The other options contain or prevent this risk at the expense of a slower or delayed sales volume.
- The product is only sold in a specific, fairly insular test market for building up experience with the product in real-life situations and containing the damage in case of problems.
- The market is entered via a launching customer. The launching customer has been involved in the final stages of the product development, has assisted in solving the teething problems, and provides confidence to other customers that the product is fit for purpose. A launching customer can be of immense value for a start-up, particularly when the product is very innovative.

3.4 Project execution plan

The project plan for developing and commercializing the innovation outlines the important issues for the technical, commercial, and business development. A key factor for success in the technical development is that all aspects of the product have to be included from the start, or, in other words, the total supply chain for bringing the product to the customer has to be taken into account in the development of the innovative product. It is almost a 'standard mistake' to start the development by focusing only on the technical performance of the product, such as maximizing output and efficiency, minimizing energy, space or labor requirements. As a result, the product may be very good, but it could still be unsuccessful because it ends up to be expensive to manufacture, or not reliable in operation, or not easy to use by the customer. It is essential to include in the product development aspects such as how to manufacture and distribute the product, the total cost picture, ease of maintenance and servicing, robustness and reliability in operation, exterior design, and selling points from the start. The case of the Super Reactor provides a striking example of a serious setback caused by a supply chain issue in stage III that was not timely recognized in stage II.

Super Reactor

Fasmall,[2] a leading technology company in the United States in the area of reactor intensification, had developed a very advanced reactor for a chemical process that could achieve the same or better performance as a conventional reactor, but at one-tenth of its size, offering great potential for cost reduction and safer operation. The test reactors were made from a special alloy and meticulously manufactured with a special, highly advanced bonding process. The reactors were extensively tested and performed according to tight design specifications.

However, in the subsequent commercialization stage, the business developers realized that the global capacity of this advanced manufacturing process was an order of magnitude smaller than the expected worldwide demand for the reactors. As a result, a new, high capacity manufacturing process had to be developed and tested, leading to a significant delay in the commercialization and a loss of confidence with some of the partners.

The story of the Super Reactor is an illustration of one of the important rules in product development:

Rule 11 – Product innovation needs to include optimization of the supply chain

IKEA is a well-known example of a company in which product development and continuous optimization of the supply chain are fully integrated. Parties in the supply chain are actively advised how to reduce cost in the manufacturing and supply of the parts and the customer is the final link of the supply chain. For IKEA, this integration is not only driven by product development considerations, because control of the supply chain is an integral part of the business model.

[2] Fasmall is not the real name of the company.

The success story of Cambridge Display Technology (CDT), a technology development company, underlines in a different way the need for being actively involved in the supply chain, even when the final product is developed by other companies.[3]

Cambridge Display Technology

CDT is the leading developer of technologies based on polymer light emitting diodes (P-OLEDs). P-OLEDs are designed for use in the latest generation of electronic displays and in lighting products, offering many advantages over the liquid crystal displays (LCDs) and plasma displays now used in flat panel televisions.

CDT was established in 1992 and has a heritage with a direct line back to the original discovery of organic electroluminescence from polymers at Cambridge University in 1989. Today CDT is part of the Sumitomo Chemical Group.

CDT supports its partners in their development of product and supply chain in all aspects, from materials and know-how through to device development and manufacture. Examples of some of these are:

- *Partnering with amongst others (a.o.) Thorn Lighting and Pilkington, CDT is working to develop a pilot line process for the fabrication of P-OLED solid state lighting devices for integration into luminaires.*
- *Working with Silvaco, Plastic Logic, Eight19, and Cambridge University, CDT aims to develop robust optical and electrical models that can predict trends in performance, ultimately to accelerate optimization of organic devices.*
- *Understanding and predicting the flow and drying behavior of ink jet printed P-OLED drops is an important step to improving manufacturability. Working with ANSYS, computational fluid dynamic modeling is being employed.*

The technical and commercial development should be an integrated effort, with emphasis on the technology in the beginning and shifting to the commercial aspects over time. The team composition and leadership will reflect these changes, but both technology and commerce need to remain actively present in the effort. The historic approach where product development and commercialization were sequential efforts, with a formal transition of the product from one discipline to another, has proved to be inefficient and a source of unnecessary failure.

It is important that after the product has been demonstrated successfully in stage II, the commercial or business development manager takes the firm lead in building the supply chain and complete the technology development. The objective should be to freeze the product and only allow changes that are indicated by feedback from users. Adequate technical performance in terms of capacity or efficiency should have been

[3] www.cdtltd.co.uk.

part of the demonstration; change may be needed to improve aspects such as simplicity of handling, cost reduction, or esthetics of the design.

In principle, a project execution plan has three main elements:

- A list of all the activities that have to be carried out with an estimate of the costs and time involved.
- A timeline chart that indicates which activities can or need to be done in parallel or sequential.
- A monitoring system for tracking project progress and expenses against budget.

Usually, the execution plan will include further details such as targets and objectives, the parties responsible for the activity, review sessions, and decision points. For a development project, the plan will need to be simple and flexible because there are no firm specifications and there will be unplanned, unexpected changes in the project. Figure 3.3 presents a simplified, schematic example of a timeline for a stage II development project. A comprehensive schedule should also include such aspects as assessment of safety, reliability, maintenance, and patent position.

Managing an innovation development project to successful completion is not simple. Often occurring challenges are discussed below.

Cost and time overruns

There are few innovations that do not take more time or are more expensive to develop than initially planned or expected. A factor of two is not atypical, but much higher figures are not uncommon either. Managing overruns by themselves is difficult

#	Activity	Party	Time
1	**Prototype I**		
1.0	Concept Design		
1.1	Technology A		
1.2	Technology B		
1.3	Assess Supply Chain		
1.4	Design I		
1.5	Construct and Test		
1.6	Review	Team	
2	**Prototype II**		
2.1	Design II		
2.2	Construct and Test		
2.3	Review	Team	
3	**Demonstration**		
3.1	Construct and Test		
3.2	Review	Team	
4	**Transition**		
4.1	Reasses business plan		
4.2	Project plan stage III		
4.3	Go-No Go decision	Board	

Figure 3.3 Product development timeline.

enough, but under conditions of continual financial constraints, it becomes a real challenge; fortunately many innovators love a good challenge with poor odds. The intrinsic conflict in the innovation quest is that planning an innovation project is both an absolute necessity and a near impossibility; problems and setbacks tend to occur at unexpected points and will upset the planned schedule.

Demonstrating the product

Demonstrating the performance of the product may look like a straightforward exercise, just a simple case of setting a number of criteria and carrying out a series of tests for meeting the criteria. However, results tend to be less unequivocal than desired. A typical problem is that the prototype is different from the real product and cannot meet all the design standards during a test or not all the time. The test unit is usually smaller or it is operated manually rather than fully automatic and this can disturb the interpretation of the results or be the reason that certain performance standards cannot be met. Another complicating factor can be 'operator error'. The product is new, and if the demonstration unit is tested by a third party, the user of the unit may have little experience and unintentionally create an upset. It is quite unusual that the demo unit meets all criteria and a degree of interpretation will be needed whether the underperformance was intrinsic or due to the impact of external conditions, and the prototype may need to be redesigned for completing the demonstration successfully.

Managing expectations

Managing expectations has to be done both externally and internally. External parties may include investors, partners, as well as early customers involved in testing the product. They all have certain expectations and want to be informed on the progress. Although they will all admit that setbacks are a natural and intrinsic part of the innovation process, communicating delays and cost overruns will never be easy. Internally, it is unavoidable that the energy level, optimism, and expectations will go up and down with the progress of the project. It is part of project management to push the project forward through the dips, but keeping the external expectations high during a dip is a challenge that is difficult to meet.

Being ready for the launch

There are two critical points in the development process that are difficult to manage and often lead to troublesome conflicts. The first one is to decide whether the demonstration has been successfully completed and the second one is deciding whether the product is ready to be launched. The first decision may lead to a conflict between technical and commercial interests, but this is a conflict that can be managed and a certain delay in completing the required facilities does not have to be critical. But this is not the case for the product launch, when a fault can be fatal. When the date for the launch has been planned and announced in advance, there will be strong pressures to adhere to the date, but if the product or the organization are not ready in all aspects, the product should not be launched. It needs a strong leader to take that decision in the face of intense pressure to proceed and many a leader will launch and hope for the best.

3.5 Financial planning

Financial planning in the traditional sense with forecasts of costs and revenues is not the core financial activity in the early stages of the company. For a start-up company, financial planning is in the first place developing an understanding of the problems and options for financing the innovation and being prepared for the unexpected. Financial planning has to find the way to get the money that is required for the innovation development at the right time and from the preferred capital providers. There are several ways in which an innovation can be financed, but for each stage, there are only one or two typical options.

In stage I, the start-up company has very little of value that can be used as collateral for a loan or that can be sold. As a result, the basic option for this stage is that it has to be financed by the innovators themselves, often with the help of family and friends. There may be some additional seed money available from an incubator, crowdfunding, or a fund that stimulates innovation by starting companies.

In later stages, the company starts having something of value that can be used for financing the innovation efforts. There are two basic sources of money — investors and partners — and two sources of value: shares and IP, monetizing the IP of the company by selling licensing rights or transferring ownership of the company by selling shares. Selling IP rights will in general be less attractive than selling shares, because with licensing rights, the company loses potential revenues and market share rather than potential profits, and this restriction to the size and liquidity of the company can be considered a higher penalty. However, a partner should bring in very valuable complementary capabilities and/or access to critical parts of the supply chain.

The 'silver rule' is to sell shares at the start of the project, because they may go down in value afterward, and monetize IP in later stages because patents increase significantly in value after demonstration and in particular when the product is commercial.

There are many varieties of providers of risk capital, but the principal representatives are business angels and venture capital companies. Besides a share in the future increase in value and profits of the business, these investors may also want a say in the way the business is run. This can be seen by the innovator as a loss of control, but involvement of investors can also bring access to expertise and business intelligence. Investors can bring more than just money to the company; they can also contribute with information and good business discipline.

Business angels are private persons who typically have extensive experience in the type of business that they are investing in and want to be involved in steering the business. Their advice, asked or unasked for, is valuable and their network can bring in customers and partners. Equity funds tend to be less interested in an active advisory role but will carefully assess the management qualities of the innovators, and if need be, change the CEO or the management team. They may also request a seat on the board of the company.

Partners and equity funders are the prime sources of finance in stages II and III. At the end of stage III, a launching customer, interested in accelerating the launch of the product, may provide additional funding, for instance against future discounts or a share in the business. Quite often parts of the development can be financed with subsidies, providing additional money to the funds that the company could raise from investors. Governments may want to stimulate innovation in certain sectors of industry, such as renewable energy, ITC, or high tech in general. Also local and regional authorities may provide subsidies for supporting desirable developments.

In stage IV, banks may enter the finance picture. The company now creates revenues that can pay for the interest and can provide collateral such as good debtors and fixed assets. Once banks start financing, the start-up phase has been passed and the innovation has been successfully completed.

It is very important to raise as much money as possible at the start, because often the value of the company will go through a depressed period and that is a poor time for raising money. The depressed period is a well-known phenomenon in innovation and generally known as the Valley of Death, because many innovations fail in this stage.

Understanding the dip in the value of a start-up company is an important factor in the financial planning. Figure 3.4 illustrates the financial mechanisms that create the Valley.

It would seem logical to assume that the value of an innovation increases 'gradually' from its Starting Value (SV)[4] to its Potential Business Value (PBV). This may be

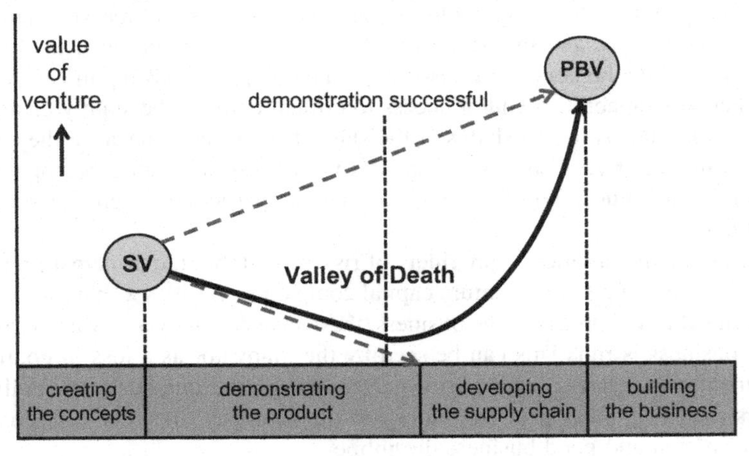

Figure 3.4 The Valley of Death.

[4] The SV is the sum of the option value and the working capital in the bank received from the investors. The PBV is the (discounted) NPV of the business at the time of launch.

the case in the eye of the innovator with inside information and an optimistic outlook, but this may not be so from an external perspective. Once the investor has provided the money and the innovation project starts, the value of the venture may well start to decrease. The cash in the company reduces and the confidence that the innovation will be successful erodes with the passing of time, because the development effort will not instantaneously lead to positive results. There will be setbacks and delays, and the time left for successful demonstration will be getting shorter. Directionally, the value of the company will go to zero at the end of stage II, when the money is gone and the demonstration of the product technically fails. Only when the innovation has been demonstrated successfully will the value of the venture increase again. The area between the curves in Figure 3.4 could be called the 'Value Space'; it represents the range in which the value of the company will fluctuate in response to good and bad news, reflecting the changes in expectations against progress made.

Innovators, often supported by their financial advisors, are inclined to base their financing strategy on the straight line pointing at the PBV and go for low amounts at the start with the expectation to raise money again on better terms when the company has increased in value. This is a high-risk strategy and not robust against setbacks and dips. It is prudent to base the financing strategy on the solid, low curve of the Valley. The overriding objective of the financing strategy for a start-up is: do not run out of cash!

The Valley of Death

The Valley of Death refers to the well-known phenomenon that many innovations fail at the end of stage II. This depressed period in the development of an innovation project can come about for a number of reasons.

- Demonstration fails

The obvious cause is that the demonstration of the product is not successful and the innovation fails. However, even if the demonstration does not fail, but is not straightforward and needs a few trials and restarts, the momentum tends to slow down and confidence can go to a low point.

- Lack of capital

Typically, the development costs are higher than expected and the company needs refinancing. However, raising new money before the product has been demonstrated is difficult because the value of the company is at a low point and the refinancing may fail.

- Dip in energy level

The end of stage II represents the transition from technical development to commercial development. This phase represents a profound change process in the organization and this will also cause a dip in the energy level and confidence in the good outcome in the team. Many difficult decisions have to be made and this will slow down the progress.

Figure 3.5 shows the dip in energy level when Shell Global Solutions had to change from an internal technical services organization to a commercial technology company. In this instance, the company recovered quickly and completed the transition successfully. However, if the dip in morale within the company extends over a longer period, it may slow down progress to such an extent that chances of success start to fade.

In an innovation quest, another dip may again occur at the launch of the product for similar reasons as for the first dip.

- The product leads to complaints from customers.
- The initial revenue streams are below expectations and causes liquidity problems.
- The supply chain and the organization are not ready or up to standards and cannot meet demand.

The last factor may well be the most typical one, but all the above factors lead to a reduction in value and energy level in the company.

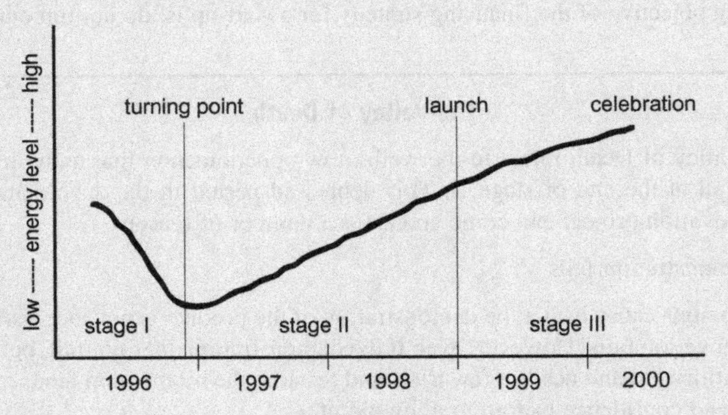

Figure 3.5 The energy dip in the Valley.

Once the demonstration stage has been completed successfully, financing will in general become easier. First of all, the chances of success have become significantly higher, the value of the patents has jumped, and the financing requirements can be better estimated. This will make it easier to attract risk capital. Secondly other forms of financing may become available in stage III. Partners will be more interested in the business and will give more money for a share of the IP. This cooperation has the additional advantage that next to money the partner also will bring in certain required capabilities and in that way further improves the chances of success.

However, to make things worse for the financial planner, the value of the company may not jump back to its 'rational' high level after the product has been demonstrated technically, but remains low. This extended downward pressure on the value of the venture may result from two mechanisms:

- Investors may argue that a product has to be demonstrated twice; the first time technically with a demonstration model or pilot plant and the second time commercially in the market place. Thus, it is not the PBV that will be used as the value target, but 'point zero' is moved from the end of stage II to the beginning of stage IV.
- The PBV as calculated and used in the business plan will have been based on the average value of the revenue streams over a period of 10-15 years. But in the early years, the actual revenue streams will be lower and not the PBV is used as the value target, but the Business Value based on the actual or expected revenues in the early years.

When both mechanisms are at work, the Valley will become extended over a much longer period as illustrated in Figure 3.6. Refinancing with risk capital from investors may become difficult if not impossible without selling out completely. In the Valley, the financial planner should look at partners and launching customers for covering the cost of developing the supply chain.

The curve in Figure 3.6 can be seen as a lowest possible value curve for a successful innovation. Usually, positive news such as the signing of an agreement with a strategic partner or launching customer will lift the curve to a higher level or at least create a peak that can assist in refinancing the company.

The dip in the value of an innovation project has important implications for the planning of the financing of a project. If possible, the financing should be arranged at a high point in the value curve and be sufficient to carry the project through the dip. This would make the project more resilient against minor misfortunes and offers the best exchange rate of money for intellectual and human capital. However, it is unusual that at the start of an innovation project sufficient capital can be arranged to cover the whole development cost. Typically, the financing is arranged in stages,

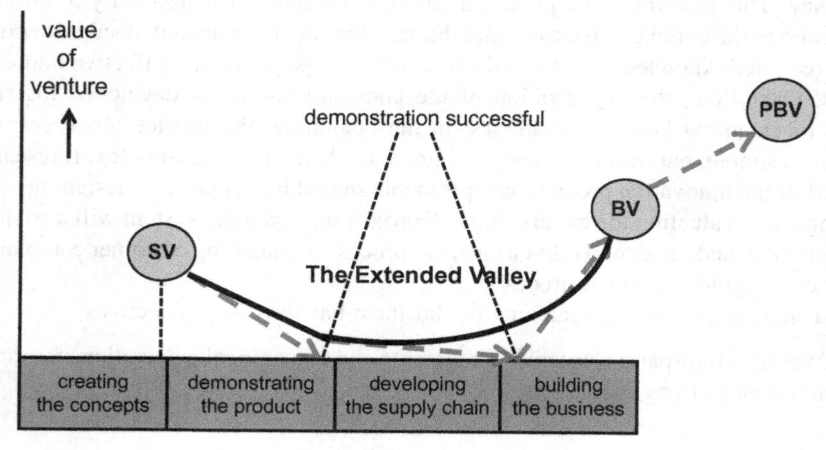

Figure 3.6 The extended Valley.

and it could be very attractive to agree that the money at the subsequent stages is provided against the same valuation as at the start. The story at the start, with the hopes and expectations it can create, is often worth more than the reality in the first years of the company. Raising money in the Valley is not impossible, but it requires additional effort. It is important to refresh the sales story for the investors with sufficient good news that can create a peak in the Valley. But in general, it will be difficult and costly to raise money in the Valley and it is important to start the effort early because it may take more time, and the problem may turn into a crisis when the money runs out before new capital has been raised, or even worse capital can no longer be raised because the value of the company has become too low. Rule 12 is a survival advice for the financial planner.

Rule 12 – Refinance before the money is needed

It may be required to create additional sources of income when the money that can be raised from investors, subsidies, or partners is not sufficient. It is not unusual that a certain amount of development work is done for partners or other third parties. This can be desirable for both parties for using the expertise in the start-up company and can be a valuable source of income in difficult times for paying the bills and keeping the team together. The danger is that it diverts effort away from the innovation project and the resulting slow down may initiate the onset of unsuccess because it extends the dip in the value curve. If it becomes a major source of income, the company may gradually change from an innovative techno-starter to a R&D company selling man-hours at a profit. One could consider the failed innovation as collateral damage, but it is also possible to consider such a development as a timely and successful change in destiny.

3.6 Organizing the business

For a small start-up company, this topic may look alien and not an issue of great urgency. This is correct to a point, but at a certain stage, it is necessary to provide structure to the company to ensure that the activities are coordinated, results are properly recorded, knowledge is consolidated, and the operations are effective and efficient. Over time, the organization of the company has to be developed together with the required business processes. In the beginning, the finance processes will be fairly simple, but over time, they will grow and become more complex. If research is part of the innovation project, then protocols should be in place for designing, executing, and evaluating the experiments. A project management system will also need to be established rather early, but a business process for handling customer complaints may come quite late in the process.

For a start-up company, organizing the business has three key objectives:

- Creating a transparent structure for steering and managing the operations for meeting the objectives and targets.

- Developing a series of protocols for executing the operations effectively and safeguarding the integrity of the data and know-how.
- Establishing systems for meeting the legal requirements.

Integrity in data and systems

For a techno-starter, the effectiveness of the development function is key to success and safeguarding the integrity of the research and development data is essential. It is important that the data created during experiments and testing are correct and fit for purpose. But this does not happen by itself. Even the results of the best scientific experiments carried out by top researchers may be biased in favor of the prevailing hypothesis and assumptions. There is always the danger that the researcher jumps to conclusions or find what is expected to be found. Thus, experiments and interpretations have to cross-checked, equipment needs to be tested for ensuring that the data generated will be consistent and reliable, and conclusions have to be challenged. A good protocol will separate the design, execution, and interpretation steps of the experiments with involvement of different persons for the various steps and ensure that the system is transparent. When the technical data are not sound, the project will fail at one stage or another, and hopefully this occurs at an early stage and not when the product has been launched into the market; at that stage, the costs are high and unrecoverable.

But not only in the development area is a system of checks and balances relevant. The need for integrity in the area of finance, contracts, and purchasing is equally important. In a starting company, the protocols tend to be not very rigorous and that represents a significant risk that is often not recognized. This risk should, however, be carefully managed in order to avoid very unpleasant outcomes (see Section 6.2).

Operating effectively and efficiently

In the initial phase, when the company is still very small, this should not be much of a problem. The team is small, the commitment is high, and the communication lines are short. With the progress of the project, the mix of the required capabilities has to change and becomes more diverse and complex, and this needs proactive management. Finding the right balance between the capabilities that need to be in-house or those that can be accessed via partners or outsourcing is not always straightforward. Core capabilities need to be in-house. The extreme cases of either all capabilities in-house or all outsourced have proved to be inadequate or inefficient.

When the start-up moves to or beyond stage II in addition to technical departments, it will need commercial and business development departments, and ultimately also supporting functions such as human resources, ITC, legal, or even PR. For a start-up with 10-15 staff, there is not much need for a formal structure and functions such as finance may still be part-time jobs. For a technology-based company, a simple functional structure for the organization is usually adequate, with departments such as R&D, Commercial, Business Development, and Finance reporting to the CEO. When the organization grows to a level of 50-100 employees, supporting departments may become required and the organization might become more complex or shift to a different model or elements of it, such as the matrix or network model.

Important features of a good organization are being flat, flexible, and low on bureaucracy and hierarchy. The business processes and internal communications have to be developed in line with the complexity of the organization for ensuring alignment and consistency. Part of organizing for success is ensuring that these requirements are met in a timely fashion.

Meeting legal requirements

It is obvious that a company should take care that it adheres to all the legal and regulatory requirements, but for a start-up this can be a challenge because the company may simply not be aware of all the rules on such diverse issues as waste handling, emissions, employment, training, reporting, etc.

It is best practice that Safety, Health and Environmental (SHE) aspects receive priority attention. An incident in the SHE area is always a serious event with high emotional elements. Great care should be taken for preventing 'avoidable incidents'. A SHE accident always leads to involvement of authorities with potentially serious negative impact on the company, including temporary or permanent interruption of the activities.

Summary I

Innovation can be described as creating a bridge between the world of knowledge and inventions, and the world of customers and value. For a start-up company, a good working definition for innovation is *"Innovation is creating a successful business from an insightful idea."* Crossing the bridge and creating a business with an innovation is a process with four distinct stages:

Stage I — Creating the product and business concepts.
Stage II — Developing and demonstrating the product.
Stage III — Developing, or creating access to, the supply chain.
Stage IV — Building the business.

The stages are separated by tollgates, and at each tollgate, the value of the innovation and the business model needs to be reassessed. Before starting the next stage, the project plan should be agreed and financed. Financing the product development stage is a critical exercise as the innovation project has to cross the Valley of Death.

Success in innovation depends to a large extent on understanding the innovation process and how to execute it. Twelve rules for success, or better for avoiding unsuccess, can guide the innovation quest of a start-up company.

1. Take 1:10 as the chance of success in innovation.
2. Think twice before you start.
3. Do not start alone.
4. Passion and perseverance are the engines for success.
5. Timing is of the essence.
6. The winner is unpredictable.
7. Involve the customer early.
8. It takes twice as long and it costs twice as much.
9. Do not start a new project before the first one is a success.
10. Even the best innovation concept is partly wrong.
11. Product innovation needs to include optimization of the supply chain.
12. Refinance before the money is needed.

Success in Innovation

Part II

Factors for Unsuccess

Most innovations fail. Managing innovation aims to improve the odds and the best way to do that is identifying and understanding the conditions and factors that tend to lead to failure. Part II describes the most important Key Unsuccess Factors (KUFs) and the ways for avoiding them or to mitigate their impact. KUFs are 'avoidable standard mistakes', and unsuccess in innovation is failure that could have been avoided.

4 Wrong Starts

4.1 Solution looking for a problem

The phrase 'solution looking for a problem' stems from the previous century when research and innovation were carried out differently. It was the time of 'technology-push' innovations resulting from inventions created in the research laboratories of large companies. Research, manufacturing, and marketing were separate silos in a company, each with their own responsibility and targets that were not necessarily aligned. At the annual coordination meeting, researchers would proudly present their clever inventions to the marketing people for further commercialization and would be disappointed when marketing did not show the required interest and proper enthusiasm. The problem was not that the inventions were not clever or creative, or that the researchers had not identified a problem that the invention was intended to solve. The researchers did have a clear problem definition, but they defined the solution from their own technological perspective rather than in terms of a value proposition from the customer perspective. The solution looking for a problem KUF represents the pitfall that the inventor assumes that his solution to a perceived customer need also is the 'solution of choice' of the target customers. The most common mistake is the assumption that a superior performance measured in pure technical terms will also be the preferred choice of the customer. Philips can provide a well-known example of the merits and drawbacks of technology-push innovation, but this case study could have been based just as easily on the experience of AT&T or any other research-led company.

The era of S&T[1] push

The large research facilities of Philips in The Netherlands, nicknamed NatLab,[2] were established in 1914 for creating a patent portfolio based on independent research. At the time, most large companies shared the view that fundamental research was essential for creating the science-technology-innovation spiral. After World War II, the NatLab became a world-class research facility under the inspired leadership of Casimir, a top scientist as well as a member of the Board of Philips. He developed a famous set of rules for research policy. He preferred leading above managing and the key to success was to give a fair degree of freedom and generous budgets to top talent. Researchers could select

[1] S&T = Science and Technology.

[2] NatLab = Natuurkundig Laboratorium (Dutch for Physics Laboratory).

their own research areas as long as they were challenging, interesting, and relevant for the future of the company. Many technological successes can be claimed from that period, such as the development of the VideoLP as predecessor of the Compact Disk and significant contributions to ITC technology. The standard for success in research was academic acclaim.

In the 1970s, when Casimir had left, the belief in fundamental research as the engine for innovation started to be questioned. It was expensive and the link between research and success in innovation became blurred. NatLab management made a series of poor decisions and the best technology was not always the commercial winner as the battle between the rival technologies for the video recorder (VHS by JVC, Betamax by Sony, and V-2000 by Philips) in the 1970s and 1980s showed. The budgets were reduced significantly over time, and the number of staff decreased from over 2000 at its peak to 600 early this century.

In 2000, Philips started to convert the NatLab into a high tech campus with research facilities open for other companies. In 2012, Philips sold the High Tech Campus Eindhoven, host to over 100 companies and 8000 researchers, to a real estate investor. The era of technology-push innovation based on closed, in-house research had come to an end to be replaced by open innovation for 'simple and sensible' products.

The story of Philips illustrates that technology-push innovation has a very successful place in the history of innovation. Ford could very successfully sell its T-Fords in the 1920s with a wide range of colors as long as it was black. The United States a century ago, Europe after WWII, the Asian Tigers at the end of the previous century, and India early this century are examples of emergent markets where technology push was successful and governments developed an active S&T policy. In emergent markets where demand is high and the choice for the customer is limited, the supplier of the product rather than the customer is in the driving seat and technology-push can be successful.

The days of pure technology-push innovation are long past in the developed world. There have been attempts at applying 'marketing-led' innovation, as used for incremental innovation, also for radical innovation. But that approach was even less successful because radical innovation is very difficult to test with traditional market research tools and it often needs a technological breakthrough invention. The term 'marketing-led' innovation may well be misleading anyway. There is a lot of 'marketing-push' innovation going on, and the success is not better than technology-push innovation. Marketing is continuously pushing 'new', improved products onto the shelves of supermarkets, but nine out of ten of those incremental innovations are withdrawn from the shelves within a few months.

However, the problem of 'solution looking for a problem' is fairly intrinsic to innovation and not easy to resolve. Innovation will often have a technological component, and although innovation should be 'technology enabled' rather than 'technology pushed', by its very nature breakthrough innovation will easily have an element of 'solution-push' in it. From that perspective, technology-push and marketing-push are just special cases of solution-push. The difficulty remains that the best solution as selected by the innovator will not necessarily be the same as the one preferred by the customers. Thus, a techno-starter today may still be in the same situation as the researcher in the grand laboratories of the previous century. The 'solution looking for a problem' KUF is an issue of all ages and that is why so many innovations fail and become part of the large collection of 'forgotten innovations'. The invention of Leopold Richard Breisach of New York City provides a classic example of 'an innovator's solution for a perceived customer problem'.

Forgotten Innovations

In 1858 Breisach applied for a patent for an invention what he called his 'rotating blast-producing chair', and also provided the required miniature model to demonstrate how such an advance on an ordinary chair could actually be used. In his application to the U.S. Patent Office, Breisach explained his invention as follows: "By rotating in your seat you will operate two bellows which will provide cool air to your head."

This 'forgotten innovation' also provides an example of why a good idea is not good enough in innovation; an idea needs to be insightful. Breisach's invention can be described as a personal, human-powered, air conditioner and that is a good idea for New York in the summer without electricity. But the innovation was not insightful, because there were not many customers that liked the solution; rotating the chair may have been somewhat distracting during dinner or conversation, and the energy required for the rotation may well have increased the temperature of the customer. Although his invention did not make money for Breisach, it did create some value; the replica of his model went for a fair amount at an auction for forgotten innovations at Christie's in Manhattan in 1996.

Most of the innovations that failed will be forgotten innovations. An illustration of the vast number of forgotten innovations can be derived from the models that had to accompany a patent application in the United States in the period from 1836 until 1880. Although many were lost over time more than 125,000 of them were sold at an auction about 100 years ago.

But not all of the forgotten innovation will have been unsuccessful. One reason could be that they were successful for only limited period of time; another reason could be that history writing focuses on successes and, in particular, on high-tech inventions. Low-technology is an underrecorded sector in the history of innovation, although it could be a rich source of inspiration. In 2009, the UK Innovation Research Centre started a project on forgotten innovations and inventions as a possible source for future innovation and creating value from past research.

There is no simple recipe for successfully avoiding the 'solution looking for a problem' issue. It is not simple or straightforward to know what the preferred solution will be of the future customer. The customer may not know it either and furthermore his preference may change over time or as a result of alternative solutions that enter the market. Finding the right solution is often based on gut feel. Many of the most successful innovations were the result of the conviction of a single person: the light bulb, T-Ford, Walkman, Facebook, I-pad. These innovations all appealed to basic needs of mankind such as mobility, entertainment, and communication, and provided solutions that were attractive and affordable. These new products were envisaged by visionaries, and the required technologies were developed to make the visions happen. In a way such innovations are 'vision-push' innovations, with market opportunities identified by vision. Market research could not have provided better responses. The key to resolving the 'solution looking for a problem' issue is being 'in touch' with your customers from the early stages onward. The experience of Fonckel, a start-up company with an advanced technology, illustrates this point. Fonckel developed a novel lighting system, Fonckel One, in close cooperation with customers. The outcome was not just a lamp but also an experience of 'touching the light' and Fonckel had a flying start.

Touching the light

Fonckel One is a lamp that can be controlled with simple gestures of the hand. The intensity, direction, and beam width can be changed by touching the exterior of the lamp. Fonckel One combines several advanced technologies, including multitouch sensing, gesture recognition algorithms, intelligent LED control, and precisely engineered optics for a fluent control of the light beams.

The unique, innovative approach adopted by Fonckel was that the interaction between user and light was designed first and the technology followed. Fonckel wanted to know from the start what the added value for the user could be. The technology makes the control of the light feel simple and intuitive as if there is no technology between the user and the light.

Fonckel One has been designed by Philip Ross during his PhD research at the Eindhoven University of Technology on the question how to control and steer intelligent products with simple gestures. The result is a lamp with a feeling that you can 'touch the light'. When Philip started to commercialize his invention, he realized that he 'did not think in such terms' and he asked his university friend Kamiel de Leur to join the company as managing director. Later on, Hans Filippini was asked to join as commercial director; Philip is still the design director.

The market for Fonckel One was tested with a pilot run of a 100 lamps sold within the network of the company and was then launched at the world's largest light exhibition, Light and Building, in Frankfurt in 2012, and it became an immediate success.

Because the technology behind Fonckel One opens up many applications, it can now be licensed under the Lumotion program, and the medical and automotive industries show a keen interest.

The Fonckel One story has many learning points as it does show that the right approach and success are linked:

- The customers came before the technology and were actively involved in the design. This is not technology-pushed, but technology-enabled and customer-led innovation as it should be done.
- Operation of the lamp is self-evident and intuitive.
- Philip Ross is a brilliant designer who recognized his lack of expertise about the commercialization process and asked partners with the right capabilities to join at an early stage.
- The potential value of the innovation was measured from the customer response already at the design stage.
- The lamp was tested in the market to check for possible shortcomings before the commercial launch.
- The business model includes more than one option for creating value: selling the product as well as licensing the technology.

Identifying a potential or latent problem is one thing; understanding the essential features of the solution is another. Generic features include simplicity, reliability, convenience, affordability, and appeal. Not all features can be built into a product because there can be an intrinsic conflict between them and more features will not necessarily lead to more value. More buttons, handles, and options also lead to more complexity, more possibilities for things to go wrong, and sometimes less convenience to the customer. Use and operation should be self-evident; simplicity of operation should be part of the design of the product. Design should be an integral part of the product innovation, not a final add-on. A good design needs to be appealing and functional. A check on the presence of complicating features should be part of the assessment of the commercial prospects of an innovation. In fact, this should be done not only for its end-use, but for the whole supply chain, for example, difficulties in manufacturing, transportation, or sales should be identified at an early stage. Packaging is a well-known example of a feature that can create problems in the supply chain and for the end-user.

A good functional assessment will improve the chance of success of an invention, but the chance of failure remains and should be part of the overall commercial assessment. It is not possible to predict which innovation will be a sure winner, but there are a few basic requirements:

- The product has to work.
- The product has to deliver what has been promised.

No product can be successful if it does not meet both requirements as the stories of Edsel and Itera, innovations that were positioned and marketed as sure winners, may serve to illustrate.

Sure winners that failed

Itera[3] was a radical new bicycle not made from metal but from molded plastic parts. It was developed by Lars Samuelsson and Jan Olsson, two previous employees from Volvo, who decided to continue the development when Volvo had halted the project. Itera was positioned as 'a friend for life', that would relieve the happy owner from the heavy pedaling on the old metal models. Itara was light, would not rust, and represented state-of-the-art technology. Revenue and sales forecasts were high, and a substantial manufacturing capacity was created in line with the high expectations.

However, the bicycles turned out to be ugly, bad, and dangerous. Very few bicycles were sold and most of them were scrapped and recycled as raw plastic. The design and materials were not good. The bicycles creaked and tweaked, but worst of all, they could fall apart at sub-zero temperatures. In Denmark, it was forbidden to ride on them.

Edsel[4] is an American car brand that was launched by Ford Motor Company in 1958. The new model was developed to establish Ford more firmly in the luxury car segment of the market. It was designed to be a radical new model and its development started under the code name of E-car for Experimental Car. The final brand name Edsel was selected out of respect for Henry Ford's son. Expectations of the public were high after a high-profile marketing campaign but turned into disinterest when the car was revealed and it turned out that it was just a regular car with a more than average number of gadgets. Sales were well below target and manufacturing was stopped in 1959. On the positive side, Edsel turned out to be one of the major learning experiences for Ford and the car industry in general.

To make sure that the novel product works, not only under controlled test conditions but also in real life, it may be useful to build in a test in the market in stage III before the real launch of the product. Although this can be an expensive exercise, it has many advantages because one gets feedback on performance from a much wider range of users and conditions than is ever possible in controlled environments. Customer feedback is very valuable and can improve the product and chances of success enormously. It would seem self-evident that a company has to listen and respond to customer feedback, but that is not an automatic given. There will be resistance in the organization to

[3] http://en.wikipedia.org/wiki/Plastic_bicycle.

[4] http://en.wikipedia.org/wiki/Edsel.

believe negative feedback on its beautiful innovation and discard it as 'noise' or 'dumbo problems'. Shell's experience[5] with an innovative, high-quality gasoline provides an example of the high cost when feedback is ignored.

Formula Shell

In 1986, Shell launched with much publicity a new gasoline on the market that promised superior performance based on a clever additive. The gasoline had been severely tested before the launch in a few representative test countries. Based on the good test results, the new gasoline called 'Formula Shell' was launched globally supported by heavy advertising: "From today not all petrol is the same." *The launch in Europe resulted in higher sales, but soon negative publicity appeared when in a very small number of cars from older models Formula Shell caused the inlet valve to burn. It took Shell, in close cooperation with motor manufacturers, a long time to find the cause of the problem and although the problem occurred only in a few countries, Formula Shell was withdrawn from the market, first in Europe and finally also in Japan where Formula Shell had been a major success.*

Upon detailed investigation, it became apparent that research and marketing had received early customer feedback about the problem in the test phase but had ignored it as 'outliers' because the problem occurred extremely infrequently.

Lessons learned included the decision to forge closer links with the motor manufacturing industry in developing new products. The loss of the brand name Formula Shell was probably the biggest loss because it had all the right connotations; the word Formula had both scientific as well as high-performance associations and was the same word in many languages.

The Formula Shell example also shows how important it is to ensure that the data used in the decision-making process are comprehensive and correct. There is a strong pressure in any development process to find the right data, that is, data that support the case, rather than reporting the data as measured, both the good and the bad.

Sometimes, the use of lead customers for testing an innovative product under even more demanding conditions than that will occur in the target markets can be considered. For example, 3M tested certain products for operating rooms in hospitals first in field hospitals in combat zones. Use of lead customers can be helpful to ensure that the technical performance is up to standard, but these users may be different from the average target customer. Thus, these tests should not replace tests in the target markets but should be an extra test. After all, it is not superior technical performance but the customer response that is key to success.

[5] *A History of Royal Dutch Shell*, vol. 3, Oxford University Press, 2007.

4.2 Too much optimism

A fair amount of optimism is one of the essential requirements for success in innovation, but too much of it is a major cause for unsuccess. There are two mechanisms that can create an overdose of optimism. The first one is that the innovator is overly impressed by the perceived qualities of his idea and the second one is the drive to make a commercially attractive business plan that will create interest with potential investors.

There are many possible sources for unfounded optimism at the start of the innovation quest. First of all, the time and costs that the development of the novel product is going to take will be underestimated. But this factor is not the most critical source of unfounded optimism, and with proper financial planning, the usual delays and setbacks can be managed. Furthermore, this source of optimism is an intrinsic part of innovation and does not fool investors or depress innovators.

The genuinely dangerous form of unfounded optimism is created by the revenue expectations. The revenue forecasts generated for the business plan tend to suffer from creative optimism on two scores: sales volumes and margins. Typically, the sales volumes are estimated from a target share of the market to be reached in a limited number of years after the launch with a more or less exponential growth profile (see Figure 4.1a). More experienced innovators will adopt a more sophisticated growth curve because they know that growth will follow an S-curve, with low sales volumes in the early years (see Figure 4.1b). The S-curve results from the fact that early sales tend to take off very slowly because most customers will prefer to wait for the

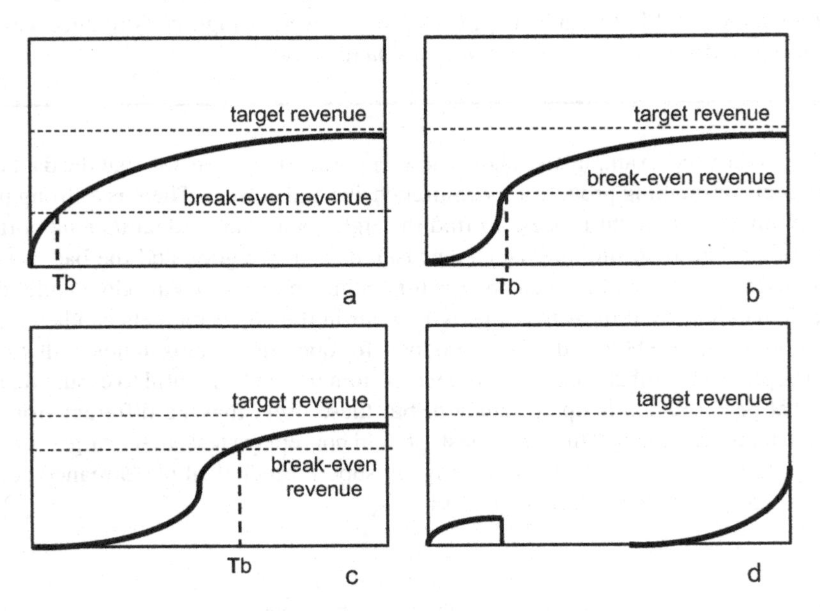

Figure 4.1 Revenue forecasts.

feedback from the early adopters. If that response is positive, the growth can be very fast until it slows down again when market saturation is approached.

For radical innovations, the start of the S-curve can be fairly long (see Figure 4.1c). Even for the cell phone, one of the most successful recent innovations with incredible high market penetration, it took many years before the sales volumes took off. Sometimes, the take-off period can last too long as in Figure 4.1d, where the initial launch is unsuccessful and only a second generation makes it to the market successfully. The key understanding is that it can take a long time before sufficient customers are willing to try a completely new product. There is a natural resistance in most markets to switch to unknown products.

However, not only the volume of sales tends to be low in the early stages, but also the margins will often be hovering around zero. Besides better performance, first-time customers will also need a cost incentive for switching to a new product, and in the early stages, the cost of sales can be high. The phenomenon illustrated in Figure 4.1d can be the fate of good, radical innovations that are launched into the market ahead if its time.

Success as second generation

QUBE, the first interactive cable television system, was launched in Columbus, Ohio, in 1977. It became quite popular, but the costs of expanding the system nationwide became too much of a financial burden for Warner Cable and, together with privacy concerns from subscribers about the use of data, led to a phase out of the QUBE box in 1984. However, many of the concepts such as pay-per-view programs, special interest cable television networks, and interactive services reemerged successfully a few decades later.

The electric battery car has the potential to become an example of a late success as second generation. A hundred years ago, the battery car was more popular, more reliable, and faster than the gasoline car. However, the gasoline car had more range and power, and ultimately replaced the electric car. These days, the electric car is trying a comeback on the back of environmental concerns.

The electric fuel cell car may even need a third attempt. The first fuel cell car was developed by Shell in the 1970s as a 'concept car' and is now a museum piece. Around the turn of the century, many car companies did development work on the fuel cell car and Mercedes-Benz announced the launch of a commercial range of models for 2002, but the technical challenges for reaching the required performance were too high. These days, there is still research going on that may enable new attempts in the next decade.

A specific form of overoptimistic expectations can result from too much idealism. There is nothing wrong with a bit of idealism in innovation; after all, innovation is all about doing things better. The reason that too much idealism can be a KUF is that it will bias the assessment of the prospects for the innovation. The danger is that the question about the potential preference of the customer is changed from 'will the customer like it' to a moral request 'the customer ought to like it', with the implicit expectation that the response should be positive. Ideals are good but not necessarily sufficient for success. Most people will have a positive feel when a novel product is 'green' and supports 'sustainability', but that may not be enough reason for buying it. Sure, it is not a good idea to innovate a novel product that is not sustainable; customers will expect that sustainability is an integral part of a good product just as quality, reliability, and convenience. But the willingness of the average customer to pay extra for 'green and sustainable' is not very robust and can fade away in bad times.

Too much idealism in innovation can become a special form of 'solution-push' innovation, where the public is expected to see the problem and the solution in the same perspective as the inventor. Such an approach can be successful when the innovation is part of a larger ideal. For example, Grassroots innovators in the United States and Europe also strive for a change in lifestyle and value system, and consequently an innovation will be a success when it serves that purpose and is adopted by the community of like-minded people. But for many customers, such a combination of a novel product and a change in lifestyle will be a hurdle.

4.3 Not the right entrepreneur

The person that has created and shaped the idea is not necessarily the right person to bring the idea to the market. It can be, but often it is not. Of course, the ideas person or inventor will bring important assets to the table such as conviction about the value of the idea and commitment to make it a success. These features are critical factors for success because hard work and perseverance can make the difference between success and failure. As Edison said, *"Innovation is 1% inspiration and 99% perspiration."*

But the one thing that innovation needs most is entrepreneurship, and more specifically, the right type of entrepreneurship, and this quality happens to be the scarcest resource in innovation. Entrepreneurship as such is not enough; not every entrepreneur is interested in or suitable for radical innovation. Most entrepreneurs just want to be self-employed for a variety of reasons, such as preferring to work alone, having problems with working fixed hours or following orders, the work can be done part-time or fits with other responsibilities, or the entrepreneur is forced into it due to lack of other employment opportunities. Other entrepreneurs are happy to do business as is, but have no desire to pursue novel opportunities. They all fit well in the world of 'Business as Usual', but not in the world of innovation. In fact, only a small minority of entrepreneurs seem to be interested in breakthrough opportunities as their preferred way of doing business, as indicated by the classification below.

Types of entrepreneurs

A study[6] in 2010 by Motivaction and $N=5$ on the attitudes of entrepreneurs in SMEs in The Netherlands distinguished five main types of entrepreneurs:

- *Expert* (23%)

Wants to work on his/her own in his field of expertise. Is dedicated and passionate, and wants to be leading in his field and continuously improve his product and business.

- *Pragmatist* (26%)

Genuine entrepreneur, action orientated, committed. Does not want to start new trends, but wants to stay abreast.

- *Custodian* (13%)

Caretaker rather than entrepreneur. Responsible but risk avoiding. Does not want to grow too big and will not pursue new opportunities.

- *Developer* (34%)

Personal development is part of drive for the business. Creative and innovative, a team player open to ideas and advice from other professionals.

- *Hunter* (4%)

Real entrepreneur, always looking for new opportunities. Convinced of his/her own vision and driven by need for success. Willing to take risks.

A similar study was done by Forbes[7] in 2011 on the entrepreneurial spirit of European executives for nurturing innovation. This study derived a similar classification with five personality types: Movers and Shakers, Experimenters, Star Pupils, Controllers, and Hangers-On.

The learning that can be derived from these classifications is that, with respect to innovation, most types of entrepreneurs will not be fit for purpose. Innovation does not just need entrepreneurship; it needs 'pioneering entrepreneurship' able to see an opportunity and driven to pursue it. This is an essential consideration for deciding who is most suitable to lead the innovation effort. Pioneering entrepreneurship should not be confused with high-risk-taking entrepreneurship. Of course, there is no innovation possible without taking high risks, but the right innovator is an entrepreneur that is comfortable in dealing with risks and also has a strong risk-adverse streak that stimulates him/her to identify risks and the ways and means of managing them.

[6] http://www.motivaction.nl/content/zes-verschillende-ondernemers-wie-ben-jij%2C1.

[7] Forbes Insights, *Nurturing Europe's Spirit of Enterprise: How Entrepreneurial Executives Mobilize Organizations to Innovate*, December 2012.

Having the most passion is not sufficient for leading an innovation project either. Passion should not be confused with competence. Entrepreneurship is the pursuit of opportunity, and passion is a crucial element in being successful in that pursuit for overcoming the setbacks and hurdles that will be encountered. Passion needs to be balanced with a clear sense of reality. Scott D. Anthony, author of The Little Black Book of Innovation, describes this balance as follows:

> *"Entrepreneurs should have to have the ability to hold two opposed ideas in their mind at the same time. They should be able to argue passionately that their ideas will change the world, and then, without changing a beat, honestly assess the risks standing in the way of its success and describe what they are doing to mitigate them."*[8]

Possessing the capabilities to handle two opposing perspectives is an important asset for an innovator. An innovator has to be both 'a dreamer and a doer', combining the ability for creating a vision with the drive for implementation. Another set of opposing qualities is being a planner with a proactive approach for tackling the challenges of the innovation quest as well as an improviser that handles the unexpected setbacks in the development with optimism and determination. For a techno-starter, a third set of qualities is important: the innovator/entrepreneur should be a good scientist/technologist/engineer as well as possess a sound commercial acumen. Since not many entrepreneurs will possess all these unique combinations, it tends to be better not to start as a single entrepreneur but as a founding team. The advantages of starting as a team are obvious: potentially there is more seed capital and the range of capabilities should be wider as well as the network and the set of experiences. Preferably one member of the team should have actual entrepreneurial experience because there is evidence that experience improves the chances of success.

[8] http://blogs.hbr.org/anthony/2012/02/dont_confuse_passion_with_competence.html.

5 Poor Execution

5.1 Not managing risks

It can offer a revealing perspective on innovation to see the whole innovation quest as a staged exercise in reducing the risks of the business proposition to an acceptable level for making investments. From that perspective, the purpose of stage II is reducing the risks on the performance of the product to a level that the investments in the supply chain can be done with the required degree of confidence, and of stage III, it is reducing the commercial risks for ensuring that the company can deliver the product to the customer as promised. In theory, the investments in the supply chain in stage III could be done concurrently with the development work in stage II on the basis of the business plan made in stage I, but the chances of success at that point typically are not better than 30% and investors will find that rather low. The effort in stage II is directed to increase the level of confidence and chances of success to a significantly higher level. At the end of stage II, chances of success in product performance should be 90% or better and at the time of launch as close to 100% as possible.

Assessment of the risks that the company may face has been done in stage I, but this has to be repeated on a regular basis, because for each stage the key risks are different and with increasing insight and knowledge the assessment may change. The first step in risk management is identifying and ranking the risks with the appreciation that each innovation project will have a series of generic and specific risks. Generic risk issues are, for instance:

- Will there be demand for the product?
- Will the technology work?
- Getting access to the supply chain.
- Financing the development and investments.
- Assessing the IP position properly.
- Getting the right mix of capabilities.

The generic risks may seem to be obvious, but they can be difficult to assess for a start-up with limited experience. In particular, the risks associated with stage III are often underestimated or ignored in the early assessments, and the problems are only recognized when they hit the company. Specific risks can be, for example, an isolated location or diverging objectives of the founders; they can be hard to identify and even harder to assess objectively. Detailed analysis of the identified risks should reveal the strengths and weaknesses of the company and should lead to a focused discussion on ways and means for mitigating the impact of these risks. A frequently used tool for assessing risks is The 25 Cell Matrix.

The 25 Cell Matrix

The 25 Cell Matrix is a simple, effective tool for qualitative risk analysis. It assesses the severity of a risk in a 5 by 5 plot with the dimensions 'likelihood of the risk' against 'consequences or impact'. There are several variations of the matrix available. Scoring can be done in different ways; with figures 1-5, or with colors green, orange, red, or in the terms as used by SmartDraw[1] as shown below.

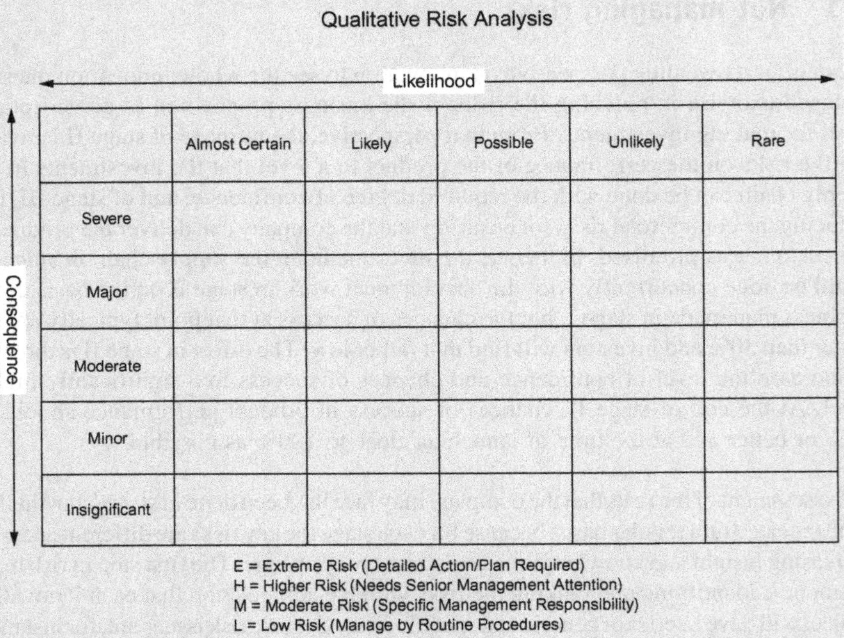

Qualitative Risk Analysis

Likelihood

Consequence

	Almost Certain	Likely	Possible	Unlikely	Rare
Severe					
Major					
Moderate					
Minor					
Insignificant					

E = Extreme Risk (Detailed Action/Plan Required)
H = Higher Risk (Needs Senior Management Attention)
M = Moderate Risk (Specific Management Responsibility)
L = Low Risk (Manage by Routine Procedures)

The matrix can easily be adjusted to specific requirements or conditions. It is a useful tool for discussing risks and developing a shared understanding of possible consequences.

It can also be useful to find out how the investors rank the risks of the project. Tackling those risks will increase the confidence of the investors and their willingness to invest.

After identification and prioritization of the risks, the next step is identifying and agreeing which actions will need to be taken and the right timing for the actions. The latter factor should not be ignored. Testing customer demand only after the product development has been completed could mean that a lot of money has been wasted if it turns out that the demand is disappointingly low. A patent search can reveal

[1] http://www.smartdraw.com/examples/view/qualitative+risk+analysis+matrix/.

whether there is IP that could block the use of the invention, and if so, discussions can be started whether the blocking technology can be licensed.

The final step in the process is the assessment of cost-effectiveness of the actions under consideration. Points to consider are that most risks cannot be eliminated but only reduced, and in general, the further the reduction, the higher the cost. It can be worthwhile to design tests for the most serious uncertainties that can be done with acceptable costs and preferably early in the project. Making a functional model or prototype and presenting it to target customers is always a good option. A more expensive test that can still be worthwhile is to test the product on a small scale in an 'insular, local' market.

When the risk assessment has been done professionally and comprehensively, the survival chance of a start-up will be considerably improved. But for a start-up company developing a breakthrough product, it will be very difficult to identify and assess all the risks. There will just not be sufficient expertise and experience and too many unknowns. Unsuccess is often the result when the assessment is not done properly and the company is hit by an unidentified or unmanaged risk.

However, all these assessments and actions relate to identifiable risks and how to manage them and mitigate their potential impact. The other contributor to this KUF is the genuinely surprise event, the 'unforeseeable' risk. This can always happen (see Section 6.3), and there are only a few options for preparing against such an event: 'Expect the Unexpected', stay lean, and build in buffers in the key activities and financial requirements.

5.2 Refinancing in the Valley

One of the most common reasons for failure is being trapped in the Valley of Death. There are many ways and mechanisms that can lead to becoming trapped in the Valley, but the three most important ones are lack of capital, losing momentum, and missing the key capabilities at the transition from stage II to stage III. These three causes are closely linked and reinforce each other's negative effect.

Not having money is the kiss of death for an innovator and running out of money in the Valley is a nightmare; it is not the best place for refinancing an innovation project because the value of the company's silverware is at a low point. Section 3.5 explains why the Valley is an intrinsic phenomenon in innovation because the value of the company can erode gradually until the product is demonstrated to work and may stay low until adequate revenues start coming in.

It will be clear that it is best to finance the whole project at the start because the value of the company at that point is boosted by a beautiful, appealing story with promises of high rewards. But more often than not the need for refinancing cannot be avoided; after all, an innovation takes twice as long and costs twice as much. A number of actions can mitigate the costs of being forced to refinance in the Valley.

- Slow down or reverse the erosion of the value of the company with good interim news for the investors. The more time that has passed and money spent without any news, the higher the chance that concerns will grow that the innovation effort may not turn out as positive as pictured. Thus, if the investors do not hear anything until the demonstration of the product has been completed successfully, the erosion will have taken place to the maximum extent. Keeping the investors up to date with progress and interim successes will help to keep the perceived value closer to the desired, 'rational' value. But erosion also occurs internally, and communicating to staff to boost morale in difficult times is an important aspect of proper project execution. Poor communication with the stakeholders can be considered a minor KUF.

- Many investors will not give all the money in one go but in tranches against achieving specific milestones. This is fair enough, but if all the tranches are issued against the starting value of the company, refinancing is 'hedged' against the downside. For the investors, refinancing in the Valley is not necessarily unattractive because they get more of the company at a low price. They will be inclined to 'follow the money' as long as they believe that the company will make it.
- Most important, do not wait with the refinancing until the point that the company is running out of money. When the money is needed badly, the negotiation position is very poor and investors may be even less willing to refinance the project. In the worst case, they will walk away or, only slightly less bad, buy all or the majority of the shares at bargain price and replace the innovator with a new CEO in whom they have more confidence. This is experiencing the Valley of Death the hard way. Surviving the Valley starts with trying to get as much money as possible at the start and asking for additional money, if need be, well before it is actually needed.

Dilution

Collateral damage from having to raise money in the Valley is that the founder's share in the company gets diluted very fast and the reward for all the efforts can become unattractively small. If new shares are issued at least equal to the preexisting price of a share, then there is no negative impact on the value owned by the founders; they just own a smaller piece of a bigger company. For the investors, the result looks better because they now own a larger share of a bigger company and have made a 'paper profit' on their earlier investment.

But, if new shares are issued below the preexisting price of a share, the value of the company goes down if the decrease in share price is larger than the increase in number of shares. In that case, the founders own a smaller piece of a smaller company. The outcome for the investors is a larger share and a 'paper loss'.

At the first issue of shares to investors at the start of stage II, an argument can be made that the founding team that created the invention should be able to keep at least half of the shares because the value of the human capital will often be higher than the value of the IP. However, after three or four share issues, the share of the founders may well be reduced to below 10%. Dilution is a painful process for the founders, but also an almost unavoidable part of building a company when external financing is required. Dilution can be seen as a growth pain and the price of success. However, if the dilution becomes too extreme, the founders may lose the incentive to succeed and this could be bad for the company and thus also for the investors.

5.3 Lacking key capabilities

It is rare that a company will have all the capabilities in-house that are required for making a success of an innovation project. The range of required capabilities is simply too wide; it may include scientific, technological, engineering, legal, financial, administrative, commercial, marketing, sales, communications, logistics, IT, and IP. A starter may have only a few of these experts in-house and has to make judgments on how to create access.

- Which capabilities are needed and at what stage?
- Which capabilities are required in-house?
- How to get access to the out-house capabilities.
- How to 'pay' for the out-house capabilities.

Reflecting about these issues also provides a good check on the viability of the adopted business model. Suppose the innovation project is based on the design of a novel, super-efficient, high-tech machine. If the business model assumes that the company will build the machines in-house and the finance manager is confident that he can secure the required capital for this capital intensive approach, then considering the required capabilities may still give an amber alert and initiate a rethink of the business model.

A number of judgments will usually be quite straightforward. The core capabilities associated with the invention or the innovative idea will need to be in-house. Legal and IP advice will be bought if and when needed. In the early stages, the need for commercial capabilities will be relatively modest. Financial expertise is essential because inadequate financing is one of the KUFs in innovation; thus, having an in-house capability is a basic requirement, but it may still be advantageous to outsource the administrative part or to pay for external advice for complex issues.

But managing the availability of, or access, to the required capabilities is not the issue in this KUF. The core problem is that the need for certain capabilities is not, or not in time, recognized. This can already happen in stage I when the business plan is prepared without business expertise. Or in stage II when the product is developed without input from the commercial or manufacturing side. The most frequent and the most serious absence of key capabilities tends to happen at the end of stage II when the company lacks the commercial and business expertise to start the development of the supply chain. Not being ready for the transition from product development to business development while residing in the Valley is a most delicate situation that can easily deteriorate into a crisis.

But even if the lacking capabilities are recognized in good time, the wrong support can be selected. It needs a certain degree of expertise to select the right provider of the required resources. Financial constraints can cause a lack of the required capabilities when in order to reduce cost the quality of the capabilities is too low. If external advice is needed, it is usually better to go for quality and accept the higher cost, because the less one knows about the issue, the more important it is that the advice is sound and can be trusted. For a start-up company, quality in advice does not have to come from expensive top consultancies, it is more important that the advisor has experience with start-ups.

The story of the Super Reactor in Section 3.4 presented a typical example of the impact of a lack of expertise, and the story of InterMan[2] provides a rather extreme example of how disastrous not recognizing the required quality of the capabilities can be.

InterMan

InterMan was a start-up company in the United Kingdom that trained experienced managers to become interim managers. The target market was skilled professionals, typically aged 45 or older, that were made redundant and found it difficult finding alternative full-time work. Potentially many of them would be a cost-effective short-term resource for medium-sized companies, constrained by tight manpower levels. It was a highly innovative approach to interim management, relying on leading Web technology.

The business model looked very attractive. The candidates would be found via the Internet, trained, and were promised at least six interviews with interested customer companies. Revenue would come from the manager, each paying £2000 for the training, and then 50% of their fees for the first 12 months. The managers targeted worked in Engineering, Finance, IT, and Operations.

Investment of over £150,000 was raised by the chairman, a very experienced and tough senior consultant, from a variety of investors. The CEO had been operating in this business space for many years and was confident that it offered scope for a successful business. The CEO handled the recruitment of companies as 'customers' and the marketing director, who used 'state-of-the-art online mechanisms', the recruitment of the candidate managers; all the other tasks were outsourced. InterMan operated in a virtual mode and had no direct office costs. The salaries of the CEO and marketing directors were capped at only £1000/month, complemented with a major share incentive scheme.

With an attractive market, experienced management team, and tight cost control what could go wrong?

The chairman accepted a full-time executive chairmanship and with no time to spend on the company left it to the CEO. The marketing director had an argument with the CEO, resigned, and vanished. It turned out that he had little relevant marketing experience and most of what he had built and done was PR related. The costs involved were over £50,000.

The CEO failed to generate more than a handful of customers willing to meet candidate managers and could not generate more than 10 jobs in 2 years. Each training course needed eight participants to be viable, but InterMan ran them with less than six on many occasions.

[2] InterMan is not the actual name of the company.

Outsourcing IT, Accounting, and Online customer recruitment turned out to be extraordinarily expensive. With such a slow increase in revenues and high cost, the company was underfunded and went bankrupt.

The key to the unsuccess of InterMan was the absence of expertise. The CEO did not really understand the business and without the steering of the chairman was not able to run the company. He did not recognize the missing capabilities and controls that were needed.

No access to the required capabilities can also be the result of stubbornness and sticking to old approaches even when being advised to change. There are innovators that are so attached to their original idea and their own ways of working that they are unwilling to change and adopt better solutions even when that implies a reduced chance on success. Unwillingness to change is one of the minor KUFs in innovation.

A key issue is always how to pay for the required capabilities. In-house resources usually need to be paid in cash, but it is quite common to pay part of the costs in shares. This has two advantageous: it keeps the cash cost down and it attracts the right people interested in innovation and willing to share in the risks. It is rare that no cash payments at all are done and the full remuneration is done with shares, but it can be an option when the lead time is short and the reward-risk ratio is high. Usually financial problems arise when the access to the supply chain and the markets has to be developed. Building a supply chain specific for the new product can be expensive and is not easily financed. Outsourcing is a good option but can be economically unattractive. Usually the best way out of this dilemma is finding a partner that can provide the access. One takes a partner to reduce the drain on working capital. The choice is between a strategic partner and a partner for convenience. The latter is usually paid with restricted licensing rights. If the partner is essential for the long-term viability of the company, the preference should be for a strategic partner and payment could also include shares. The strategy of the two companies should be aligned for creating a lasting win-win situation. The right partnership should make the company stronger, more robust, and competitive.

5.4 Losing the momentum

'Time is money' holds in business and thus also in innovation. Time erodes the value of an innovative idea, and delays in the development will accelerate the erosion. But for innovation it is also true that 'time needs money' because costs continue during delays and revenue streams are absent. A significant loss of momentum reduces the chances of success in innovation because it increases the need for refinancing and, even more importantly, lack of progress reduces the price investors are willing to pay.

There are many reasons that can cause delays and slow downs, and many are unavoidable. Negotiations with partners and investors may take longer than planned or the development meets unexpected problems. Finding and agreeing with partners or investors does take time; 'time needs money' is relevant, but 'money needs time' is also true for innovation. Unavoidable delays need 'damage limitation', and lengthy negotiations about money, rights, and obligations should be actively managed. Loss of momentum in negotiations can be a cause for unsuccess as the following story illustrates.

Fading interest

CorTech[3] had developed and patented a combination of breakthrough technologies for preventing and monitoring corrosion in the process industry. Corrosion is a major problem in that industry and a multibillion dollar business. The inventions created keen interest in the process industry itself as well in the corrosion services industries.

A major player in the service industry became interested in adding corrosion prevention to their portfolio of services, and discussions on jointly commercializing the technologies were started. Progress was slow and intermittent, but finally due diligence was carried out and Servco made an offer to CorTech. Effectively the offer was a low bid for purchasing CorTech. CorTech responded positively to the offer but with a request for a higher price. Time passed without any reaction, but finally the message came that there was no longer interest in the deal.

In retrospect, CorTech realized that their negotiation strategy had been suboptimal. Rather than entering into an exclusive negotiation with a limit in time, they had adopted a nonexclusive, open-ended approach, because they wanted to maintain freedom of action for discussions with other interested parties.

They found out that without the pressure of time there never was a focused effort and a leading party in Servco, and pressing problems of the existing businesses continued to receive priority. Over time, the deal disappeared from the radar screen of the company management team, and in the end, a change in senior management created the final loss of interest.

Besides unavoidable delays that need to be managed, there are a number of avoidable delays that can cause damage and need to be prevented. One such avoidable delay is the tendency of ambitious researchers to 'recycle the solution'. Once they have developed the product that offers a good solution to the problem, they also have ideas to do it better. And here 'better' is the enemy of 'good' because it increases the time to market and the chances are that 'better' reaches the market too late. It is better to bring Mark I to the market first and follow up with Mark II, assuming that Mark I is good enough. Another factor that can endanger the value of the innovation with time

[3] CorTech and Servco are not the actual names of the companies.

passing by is competition. Chances are that competitors are developing similar ideas or alternative solutions, because innovations and inventions are seldom unique in time. Usually customer problems and societal needs are widely recognized, and creative minds all over the world will start thinking about possible solutions. Most solutions are 'technology enabled', and when a new technology emerges, the creative minds will try to apply it. It may be a moot point whether being first is always best; it is certain that being late in the game is not good.

Another not untypical cause for avoidable delay is unrealistic expectations or 'making the idea bigger than one can handle'. There is a danger that a very good idea is unrealistically overvalued by the innovator and creates dreams of changing the world. Obviously, such an idea is worth a lot and the inventor will be inclined to ask big money from potential partners and investors. Even if the idea has the potential to grow very big, investors may be doubtful that a start-up can create that value. As a result, it may be impossible or take too long to get money and it would have been better to start with a more modest and credible proposition. Peter Drucker once wrote: *"Effective innovations start small. They are not grandiose."*

Furthermore, in conservative, traditional markets, it can take a long time before a radical innovation gets widely adopted and start bringing in revenues. The size of an innovation project should be in line with the capabilities of the founders.

Very serious delays can be caused by the time it can take to come to agreement in multiparty collaboration. This can be particularly serious when the parties come from different sectors — industry, university, government — and the more parties are involved, the harder it is to agree on the requirements and the road to market. Management by committee can kill an innovation as the story about TRIGAT illustrates.

TRIGAT

During the 1980s, NATO forces deployed a wide range of wire-guided antitank missiles with names like TOW, MILAN, and HOT. They had 'Semi-Automatic Command to Line of Sight' (SACLOS) guidance and were highly effective because the missile operator just had to keep the target in his cross hairs and the system did the rest. They worked by having a flare in the base of the missile that was seen by an infrared (IR) sensor in the tracker.

As is often the case, however, countermeasures were developed, and in the mid-1980s, Warsaw Pact-armored vehicles started to be seen with flare pots on their roofs. Such flare pots would confuse the IR sensor by emitting a broad spectrum signal so the tracker could not tell which was the missile flare and which the decoy. The use of such simple decoy flares threatened to completely undermine the effectiveness of SACLOS missile systems. NATO was in a difficult position and so tried various counter-countermeasures, but the real solution had to be in a completely new guidance system.

Consequently, TRIGAT was born. In 1988, France, Germany, and the United Kingdom signed Memoranda of Understanding for the laser-guided TRIGAT missile development. Belgium and The Netherlands joined the following year. Laser guidance was seen as the most likely solution. The TRIGAT program took a very long time to reach agreement on the military requirements and the discussion ran into the mid-1990s.

The commercial consortium set up to develop the system was highly complex. Euromissile Dynamics group was a consortium led by the German EADS, but it contained Aerospatial Matra of France and DaimlerChrysler of Germany. At the same time, MBDA was being formed from BAe Systems, Alenia, and others and was also incorporated into the consortium. By 1996 the Italians had joined the program and the missile was due in service by 2002.

Toward the end of the 1990s, it became clear that there were much better systems emerging from the United States based on the use of advanced image recognition capability. Chief among them was the Javelin which was the world's first truly man-portable 'Fire and Forget' system. With Javelin, the operator just has to lock the missile onto the target; it then recognizes the target image and tracks it if it moves during flight. The operator can thus launch the missile and then take cover; the missile does the rest autonomously. This was such a huge improvement over the TRIGAT laser-guidance system that the United Kingdom withdrew from the program in 2000 and bought the U.S. Javelin instead. The other nations also started to withdraw and the TRIGAT program collapsed by 2002.

Many of the avoidable delays are the result of poor planning, and as a result, the tools, people, skills, or money are not available when needed and progress slows down. And when progress slows down, the morale also tends to go down accelerating the negative spiral. The confidence and expectation that the innovation effort will be successful goes down, and that loss of confidence may radiate to external parties such as investors and partners. Losing momentum is particularly dangerous when it happens in the Valley. The need for refinancing will increase, and escape from the Valley can become in danger.

5.5 Making wrong assessments

Unfortunately, wrong assessments are an intrinsic part of innovation. Investors will admit that they often make the wrong assessment of a business proposal. Innovators have to accept that 'even the best innovation is partly wrong'. It is rare that the business and the product in stage IV are exactly the same as the concepts developed in stage I. However, this KUF is not about the adjustments that have to be made to the assessment of an innovation in response to new data and changing circumstances. This KUF relates to those seriously wrong assessments that can cause failure or lead to opportunities missed or rewards not received.

Receiving the right reward for an innovation is not simple, and mistakes in the assessment of the value of the innovation are easily made. Great innovators do not always get the fair reward for all their efforts as the story about QWERTY and the typewriter shows.

QWERTY and the typewriter

Christopher Sholes was an American mechanical engineer and publisher in Wisconsin, USA. He invented the first practical modern typewriter in 1866, with the financial and technical support of his business partners Samuel Soule and Carlos Glidden. Five years, dozens of experiments, and two patents later, Sholes and his associates produced a working model similar to today's typewriters.

The Sholes typewriter had a type-bar system and the universal keyboard was the machine's novelty. However, this layout caused many of the most frequently used keys to jam. A business associate, James Densmore, suggested to separate the most frequently used keys to slow down the typing speed. This became today's standard "QWERTY" keyboard.

Christopher Sholes had little interest in commercializing the invention and decided to sell the rights to the typewriter for just $12,000 to James Densmore, who had convinced him that the invention was not all that significant.

Densmore in turn signed an agreement with Remington to market the device. The first "Sholes & Glidden Typewriter" was offered for sale in 1874 but was not an instant success. A few years later, improvements made by Remington engineers gave the typewriter machine its market appeal and sales skyrocketed. Typewriters became the standard office tool and today 'everybody' uses QWERTY keyboards.

Not only inventors without commercial interests but also seasoned businessmen can easily overlook the potential of breakthrough innovations. Olivetti missed the value of ARM (Acorn RISC Machine) completely when they took over Acorn in 1986. The deal more or less destroyed Acorn (see Section 7.3), but ARM, a joint venture between Acorn and Apple, was ignored by Olivetti and 'sold' at a bargain price (effectively a reverse takeover by the remains of Acorn). Now ARM, a computer instruction chip, is used in almost all of the billion mobile phones sold each year and used extensively in consumer electronics, including tablets, digital media and music players, handheld game consoles, calculators, and computer peripherals such as hard drives and routers.

For avoiding missed opportunities, it is important to be both realistic and flexible, with an open eye for changing circumstances and alternative options. Chances are that the innovative product will never reach the target market; thus, being aware of possible exit strategies that can be used to cut the losses or go for the small prize if the big prize is out of reach, is not a waste of time.

The standard alternative business options such as licensing, selling the business, or changing to a niche market should be part of the business plan. Licensing the technology rather than bringing the product to market can be considered once the product has been demonstrated and can be an attractive option if the investments in the supply chain are high and risky. The drawback is that the product has not yet been proved in practice and that will reduce the license fees that can be demanded initially and can make it difficult to find a first customer.

Selling the company once the business is running, is often seen as an attractive opportunity for start-up companies. Selling to a large company can be a win-win situation for both parties. The condition required for a win-win situation is that the large company with its financial resources and a better supply chain can expand and increase sales more rapidly than the start-up, and thus can pay more than the intrinsic value of the start-up company.

However, assessing these standard opportunities at the typical times may not be good enough. Considering the standard options at the 'wrong' time may also lead to identifying opportunities; for instance, selling the company in stage II can serve as a 'stop-loss' exit strategy. An unusual, but attractive, opportunity for a start-up (in a less successful phase) is to buy or merge with another small company (in trouble) with a product that combined with the own product can create a new, more exciting product. Product-service combinations can be much more profitable than the sum of the separate propositions. It can have unexpected benefits to look at the research and product data from a different perspective and check if they show a potential for another innovative product with a different set of features. This requires an open mind and a willingness for changing destiny.

6 Unhappy Endings

6.1 Not keeping costs down

Once the product has been launched into the market, the mood in the start-up company changes and the same excitement as at the start returns. With a positive response from the market and a promising growth in revenues, the objectives and style in management and operations may change. Time has come to reap the rewards of all the years of hard work; salaries can increase and bonuses may shift from shares to cash. The organization has to grow in line with expected demand in the markets and possibly time has come to move the business premises to a better location; from a modest cost science park or industrial area to an office in the city center closer to customers and consultants. However, this happy picture should also be a cause for serious concern.

The consequence of all these 'logical' changes is that costs will start to increase rapidly, but if the revenue growth is less than expected, the drain on working capital is fast. That costs grow faster than revenues is fairly typical in the start-up phase because revenues increase slowly at the beginning of the S-curve and costs can easily grow linearly with time or worse. Whereas operating at a loss may be unavoidable in the post-launch period, it should not lead to liquidity problems because it is a difficult time for getting money. Investors are not interested to invest more money but are looking for dividends, and banks prefer to wait until the revenues have grown and become more robust. Thus, management should take timely measures for maintaining adequate liquidity and keeping costs down. Frugality has to remain a core value at least until the revenues and margins are adequate. The two key messages for a start-up in the early phase after the launch are:

- Stay lean and mean.
- Cash is king.

Not keeping costs down can also lead to another mode of unsuccess. When the working capital runs out and it is difficult to get money from the bank, the value of the company plummets and the company can become an easy target for bargain hunters. Young companies that have brought a new product successfully to the market are always in the picture by large companies as part of their innovation strategy. In itself that is a good thing and a desired result for many innovators as long the company is bought at a fair price, but it is a form of unsuccess if it is sold undervalued as a result of poor management.

Of course, undervaluation can also occur with the best cost management. Sales in the early years can stay low for many reasons and the confidence that the business will be successful may erode and that will impact on the value. The reasons for low sales can be many and diverse, ranging from those that can be influenced, such as a limited number of points of sale or the product needs to be improved, to factors that are beyond control, such as recession, competitive products, market being more conservative than expected, or the timing of the launch was, in retrospect, wrong.

6.2 Misinformation and deceit

This chapter could also have been called 'Lack of integrity' or 'You have been cheated'. Both titles refer to the two risks that start-ups are particularly exposed to. One is not having the correct information and the other is misuse of information. Start-up companies are more at risk than mature companies because the formal structures and protocols to safeguard data integrity are not yet in place or fully developed, and the team tends to work on mutual trust and shared objectives. Deceit can take many different forms, ranging from data that appear to be unreliable to money that disappeared. The deceit can be unintended and just the result of substandard work or the outcome of a devious plan. Cheating as factor for failure tends to suffer from benign neglect in innovation, and therefore, this KUF happens more frequently than needed.

This KUF is not about a disappointing deal or financial result. Unexpected, unpleasant monetary outcomes are almost unavoidable in innovation because shares and IP are exchanged for money, and the fair value of IP and shares is notoriously hard to determine during certain parts of an innovation process.

A not uncommon form of overvaluation relates to patents resulting from academic type research. The term 'academic type research' has been used here to indicate research that has been done at universities or research institutes and has continued beyond the fundamental stage but not much beyond the early stages of practical development. At that stage, it is possible to do experiments in the laboratory that show that an invention does work and can be used for applying for a patent. However, the gap between the invention and the ultimate product is still substantial. The difference in value of a patent on its own and the company based on the invention can be compared to the difference in value between a rough and a cut diamond. Just as a diamond needs an expert cutter to create the brilliance with that special customer appeal, does a patent need an entrepreneur for creating the product with the real value. Furthermore, a patent tries to hide as much as possible and claims are made beyond the tested conditions. Because of this potential gap between theory and practice, or claims and actual performance, patents bought or licensed by spin-offs from universities can easily be overvalued.

If the research is not done properly and below standard, and as a result the invention does not work, one may speak of a form of misinformation or deceit. This form of deceit is very hard to detect because research is a self-controlling activity and external checks on the result are not readily available. The misinformation can be intentionally to make the results more attractive, or unintentionally and resulting from sloppiness and unprofessionalism, as illustrated by the story from Redcoast.[1]

Redcoast

Redcoast, an investment company based on The Antilles, intended to invest in and commercialize a process for recovering valuable metals from waste materials that had been developed by a visiting researcher at a university in The Netherlands. Several patents had been granted on the innovative process and the design philosophy was supported by two respected professors.

Redcoast charged an international process design company to make a cost estimate based on a conceptual process design, and this study, combined with data on feedstock availability, market demand, and world prices for the metals showed attractive economics. Once a partner was found that was interested in processing the waste material and operating a plant, it was decided to continue the development with the detailed design and construction of the facility.

However, the detailed process design made with an advanced computer program showed conflicting results. The process conditions selected on the basis of the thermodynamic data were significantly different from the conditions in the laboratory experiments and this discrepancy could not be explained.

A new set of experiments was designed and carried out in an independent laboratory to check the original process design data. It appeared that the overall design concept was sound, but many of the process conditions were incorrect or even impossible and the process according to the original design would not work.

Because the researcher from the university refrained from giving access to the data on the original experiments, it was not possible to establish whether the deceit was the result of sloppiness, incompetence, or worse.

Another form of deceit in start-up companies is misuse of information or creating misleading information. There are many ways to do that, and a not uncommon format is that the original inventor and founder with limited interest in the financial side is misled by the commercial partner who cooked the books in such a way that the ownership shifted from starter to partner. Misinformation and inadequate controls are a poisonous combination that can endanger promising start-up companies. The story of Haolam[2] is an intriguing, although fairly unusual, example of this killing combination because it involved the CEO.

[1] Redcoast is not the actual name of the company.

[2] Haolam was not the actual name of the company.

Haolam

Haolam was an Internet business in the United Kingdom seeking to attract very large numbers of young people across the world to tell them about their academic and career opportunities in as many countries as possible. Four investors all had a seat on the Board, although only the CEO was an executive of the company. The Board heard the messages from the CEO that it wanted to hear:

- *User numbers were rising steadily and becoming significant.*
- *Key agencies and revenue generating customers in target countries were signing up.*
- *Costs were being kept down.*
- *A new-generation Web site was almost ready and the Board was shown how impressive it would look.*
- *There were no significant creditors.*
- *The annual accounts were presented by the CEO, approved, and signed.*

Then the CEO initiated negotiations to replace the existing share incentive scheme with one that would give him and key management more shares at better terms. These negotiations collapsed and acrimony prevailed, a key executive resigned, and the CEO refused to cooperate with the Board and eventually was sacked. When the CEO had left the company, it turned out that:

- *User numbers were far lower than had been reported to the Board.*
- *Virtually no sales contracts had been secured in the past year and there were none on the horizon.*
- *The sales-driven strategy that the CEO supposedly was pursuing was a fiction only seen in the Boardroom.*
- *No evidence of the new Web site could be found on the server; it either had never existed or else had been removed by the executives in charge.*
- *Creditors had been significantly understated including fiscal debts, and the company was in fact insolvent.*

The overall result was that Haolam had hugely overspent on its new Web site and had nothing to show for it. The basic business model was reviewed by consultants and verified as sound, but given the need to reinvest substantial additional amounts, the investors all preferred to put the company into liquidation and accept the losses.

The Board had relied on the integrity of its CEO, as he was a coinvestor and a most impressive and charismatic figure. The Board did not feel the need to continually second guess the CEO, but trusted him to a significant degree. Because the CEO disappeared from the radar screen, it is not possible to establish whether greed and revenge or sheer incompetence was the root cause or driver for his actions.

In the Haolam case, a learning point is that having only one executive on the Board is risky and the financial and product development controls were too relaxed. If start-up companies would be aware for the need of adequate controls within the organization once it starts to grow and would built in the required checks and balances, failure by deceit would only be one of the minor KUFs.

6.3 Bad luck

There are cases where an innovation fails despite all KUFs having been skillfully avoided or managed. There is a degree of randomness in the outcome of an innovation quest and even the best prepared and executed quests can fail. It is said that Napoleon selected his generals based on whether they were lucky, knowing that in such complex games as wars and battles decisions have to be made with inadequate information and a leader has to be lucky in making the right decisions for achieving victory.

It is also said that one can create one's own luck by playing the game with excellence, based on experience, skill, passion, and determination. But for winning the game, all parts have to fall into the right place, and there are external events that can ruin the best of efforts.

Chris Curry, one of the founders of Acorn Computers, recalls how a financial crash killed the commercial prospects of Keyline, a first teleshopping system from the pre-Internet era[3]:

> *"It got to the point of launch and we hit something called Black Monday. John Major's folly destroyed everything that was being created at the time. He decided that we had to stay in the ERM and the only way was to keep putting interest rates up, and he put them up until some people were paying 25% interest. Lots of people were paying 18%, but some people were paying 25%. If you are a venture capitalist, your model is that you normally expect to make a 10% return on investment. If you can put your money in the bank and get 15% without any risk at all, you stop putting it into ventures that have risk attached, and that's what happened. The entire venture capital market disappeared shortly before Black Monday and he destroyed all new growth businesses, and he didn't even notice it, of course, because he didn't have the sense to see it."*

Accidents, external disasters, and other unfortunate events can be seen as examples of 'bad luck', but it is not always a proper excuse for failure. Most bad events do send signals that should serve as warnings and for taking action. Proper preparations may not stop the negative impact, but at least could reduce and contain it. Being alert and prepared for the unexpected is an important asset in innovation.

Sometimes the accident that killed the innovation is intrinsic to the innovation and in a way 'waiting to happen'. The fire of the airship Hindenburg killed the future of the

[3] http://elleeseymour.com/2012/02/24/how-olivetti-stitched-up-acorn.

zeppelins, but using such a flammable gas as hydrogen as the lifting gas is intrinsically unsafe.

The once very popular electric car fell out of grace when during World War I the taxis from Paris were chartered for transporting the troops to the frontline but failed to reach the destination because their batteries ran down too soon. This shortcoming of the battery became unacceptable in a critical situation.

Barthélémy Thimonnier from Paris invented the sewing machine in 1830. Under unfortunate circumstances, his company went bankrupt. He continued to develop a much improved version and by 1851 the model was ready. Lacking money, he decided to go to the world exhibition in London for registering his invention and finding venture capital. Unfortunately, he was delayed during the trip, a not uncommon event in those days, and arrived two days too late. The opportunity was gone, and in the same year, Isaac Singer patented his sewing machine.[4]

But sometimes even very small events that will never be identified as a possible cause of failure can lead to a negative outcome as the amazing story of the Namibian beetle shows.

Biomimic beetle technology

This case took place around 2000 and concerned the attempt to develop civilian applications of a military biomimetic technology based on the Stenocara beetle. This beetle, a native of the Namibian dunes, collects drinking water from fog by using a unique combination of water repelling and water attracting characteristics of the wax material coating on its back. The beetle arches its back into the fog and small water droplets collect and run down from its back into the mouth.[5]

The UK military research sought to replicate the beetle's material into large sheets that could be deployed by troops operating in desert conditions. The hope was that in certain winds these sheets could trap significant amounts water which could be used for drinking. The use by Special Forces operating deep in the desert was thought to be the best prospect. The unreliability of the water yield was problematical, as it was dependent on fog occurring frequently with a particular droplet size. Although this does occur in the Namibian dunes, it is not common in deserts.

However, there were other potential civil applications such as the use of the material in tents and building coverings for collecting drinking water from humid air. The potential humanitarian benefit of this technology was recognized by a major energy company. A budget to transfer the technology was requested, a workshop was organized to create the project proposal, and the coinventor of the material was invited to speak.

[4] P. van Dooren, *En toch werkt het—uitvindingen waar je van had moeten horen*, Davidsfonds, 2012.

[5] *Nature*, 414, 33-34, 2001.

At this point, the unusual random factors started to intervene. The first one was a reluctance of the coinventor to attend the workshop; when pressed it transpired that every day he collected his children from school and refused to make alternative arrangements or to travel. The next factor was the disinterest of the management of the technology company to find an alternative solution.

Without the attendance of the research institute as technology provider, it was impossible to discuss the technical merits of the project. The requested budget for the project could not be approved and the innovation effort was discontinued.

An innovation project is a complex endeavor where many things have to turn out right and where not all factors for success are under control. Every successful innovation will have had one or more narrow escapes. Radical innovation needs a degree of good luck to reach the desired result. Having a good idea and executing the innovation process faultlessly are sometimes just not good enough for being successful. Fortunately, on the other side of bad luck is serendipity (see Section 8.6) as a form of good luck that can make the difference.

Summary II

Managing innovation for success aims to improve the odds and the best way to do that is identifying and understanding the conditions and contributors that tend to lead to failure. Key Unsuccess Factors (KUFs) are 'avoidable standard mistakes' and unsuccess in innovation is a failure that could have been avoided. The ten major KUFs are as follows:

1. **Solution looking for a problem** — The innovation is clever but does not solve the problem the way the customer would have liked it.
2. **Too much optimism** — Revenue forecasts in the business plan tend to be too high, particularly for the first years after the launch.
3. **Not the right entrepreneur** — Most entrepreneurs prefer to operate in a 'Business as Usual' environment; innovation needs pioneering entrepreneurship.
4. **Not managing risks** — Innovation is risky, and in a way, the whole innovation process is about reducing risks to an acceptable level.
5. **Refinancing in the Valley** — In the development stage, the value of a start-up tends to go through an often extended dip and that makes refinancing a challenging exercise.
6. **Lacking key capabilities** — The start-up has to shift from technical to commercial development and is not timely prepared for the transition.
7. **Losing the momentum** — Delays are unavoidable in innovation, but lengthy delays that are not managed properly will erode the value of the innovation.
8. **Not keeping costs down** — In the early period after the launch, costs tend to go up ahead of the increase in revenue.
9. **Misinformation and deceit** — Start-ups are exposed to deceit because the control systems are still immature.
10. **Bad luck** — Success in innovation needs a degree of good luck, but there is bad luck that could have been avoided.

Success in Innovation

Part III

Factors for Success

Innovation cannot be done in splendid isolation but needs support from and interaction with a diverse range of parties. Understanding the innovation arena and the actors is one of the requirements for success. Part III describes some of these factors, both the internal ones for which there is a degree of strategic choice and the external factors that are part of the innovation environment. Factors for success are favorable, supportive circumstances and can be instruments, conditions, agents, or platforms. However, the factors for success are not only positive influences; they can have negative side effects as well. Part III describes the two sides of the Key Success Factors (KSFs).

7 Internal Factors

7.1 Patents

For many innovators, a patent will be welcomed as an important tangible asset and an essential part of the monetization process of the invention. There are three ways for creating value with a patent:

1. Creating a degree of exclusivity for the product to defend higher margins.
2. Supporting licensing activities.
3. Facilitating raising money by making the IP a company asset.

The first point is often considered to represent the intrinsic purpose of a patent, but for a start-up company, the last point is usually the most important one.

A patent is granted for new and inventive technical solutions (products, machines, processes) and the purpose of a patent is to create a defense mechanism for safeguarding the potential value of an invention. A patent aims to exclude the competition from using the invention. In this way, a patent can create a competitive advantage by providing exclusivity to the patent owner and exclusivity can be used for defending higher profit margins.

A patent does not have an intrinsic value; it can only acquire value if there are customers that are interested in the product derived from the invention and if there are other companies interested in using the invention for that purpose. A patent with no product on the market associated with it or where commercialization is still far away will have little or no value. If the invention represents a real breakthrough that can open up a whole new field of potential applications, then a base patent with wide-ranging claims can become extremely valuable. Unfortunately, this type of patents represents the exception rather than the rule. Most patents have little value; in fact the average value of a patent is only a few thousand dollars. On the other side, Google valued the patents and in-house technologies of Motorola at $5.5 billion when they bought the company.

An innovator must be aware that a patent does not give the patentee the right to use the invention. The patent right is only a right to exclude others from using the patentee's invention, but it can still be possible that applying the new technology infringes the patent right of another party. To reduce the risk that one may not use its own invention, a freedom of action study is advised.

One must always consider not applying for a patent and keeping the invention as a secret. A patent is one means of defense, but keeping the technology secret is an alternative means of protecting the technology. A patent application will publish details of the invention and this may give leads to the competition for initiating research efforts to copy the invention or develop alternative approaches and circumvent the patent. Successful circumvention in turn can erode the initial advantage of the patentee. Copying will have to be challenged by means of the patent right. The secrecy approach may be taken if the invention cannot easily be derived from the innovative product. The secret recipe of Coca Cola is a well-known example of the benefits of the secrecy option.

If the technology is not patented and kept secret, one runs the risk that another party patents the technology. The technology owner may in such a situation rely on prior user rights to continue using its invention. However, prior user rights are typically limited to a country and sometimes to a certain production capacity. Thus such a situation may seriously hinder growth and licensing opportunities. To avoid third parties patenting the invention, one could publish the details of the technology to create prior art and avoid third parties obtaining valid and relevant patent rights. Larger companies often use a patent application filed in only one country as a publication strategy. This also leaves the possibility to file in other countries within one year from filing the 'strategic publication application' when it appears that the technology has become more relevant.

A limitation in the effectiveness of a patent can be that infringement can be hard to detect, and if the infringer does not agree and is not willing to take a license but brings the case to court, the legal costs can be substantial and the proceedings can take a very long time. Infringement cases can lead to high legal costs, but it can also lead to licensing income if a deal can be made with the infringer before litigation starts. In certain situations, the company who is challenging is better off with a license in the short term, rather than with a court ruling that the patent is invalid after a long legal process of several years.

On the other hand, an infringement case against the competition can be a powerful instrument for hampering the development of that business because investors, partners, or customers may be reluctant to do business as long as the infringement case is in court.

Sometimes a patent is used for defensive purposes. In this application, the owner of the patent has no interest in using the invention but wants to prevent the competition from using a specific process or method. For a start-up company, defensive patents are usually of no interest; defensive patents illustrate that for the purpose of margin protection patents tend to be a big company game.

Patents are usually required for supporting a licensing effort. It provides licensees with confidence that the technology is unique, can create more value than competing technologies, and is worth paying a fee for the right to use the invention. Licensing can offer a low cost, low risk option for creating, occasionally substantial, value from patents. For most companies, licensing will be a secondary source of income, but for a technology company, it is the core business. In that case, the patents preferably should cover a breakthrough technology that offers a significant advantage in a substantial market; otherwise the licensing income could be rather modest. A disadvantage for a start-up company is that the licensing income streams from new technologies tend to have long lead times and to grow slowly.

As a side business, licensing can be quite attractive. For example, many universities create an income stream from selling patents and licensing the rights of in-house inventions to companies ranging from spin-off companies to multinationals. A well-known example of a substantial licensing income comes from Stanford University that started their Office of Technology Licensing in 1970 that fostered many of the Silicon Valley companies, held the patents for companies such as Google and Hewlett-Packard, and earned over the years more than $300 million.

Patents can also be used to raise money, but this opportunity tends to be fairly uncommon for start-ups. A 'sell and lease back' construction can be used when there is a substantial patent portfolio generating fairly steady revenues. A bank could be interested in owning such a portfolio and the company can raise much needed capital.

All the above options for making money with patents are interesting, but for a start-up company, the dominant reason for applying for a patent is not so much making money or protection against competition, but because it is required for starting discussions with potential investors or partners. Patents makes the company the economic owner of the IP rather than the inventor, and an investor wants the security that when staff leave the company the IP is not lost as may happen under a 'keep it secret' approach. The patent right makes the IP a tangible asset and in this way can serve as a security for the investment. A granted patent will furthermore provide evidence that the innovative technology is unique and not already on the market. For this reason, an IP due diligence is often part of an investment procedure.

A start-up company should be aware of the cost factors associated with a patent.

- Applying for a patent can become expensive, in particular when the number of countries applied for becomes large. An initial application may cost $5000-10,000 and over the lifetime of the patent cost may amount to $10,000-30,000 per country.
- Infringement of the patent by competitors can be hard to detect and, if detected, can easily lead to complex disputes.
- If the patent is challenged by another company, the defense can become very expensive and the legal processes can drag on for a long time.

These apparent risks should be dealt with by means of a pragmatic strategy wherein the balance between expenditure and benefits of the patent rights is continuously monitored and the option to stop, limit, or abandon certain patents in a timely fashion is kept under review. Therefore, the patent advisor should understand the company strategy and his advice should take this into account.

7.2 Investors

Obviously investors are high on the list of important KSFs. There are two challenges in the relationship: the first one is finding the right investors and convincing them to invest in the company and the second one is to keep them happy. Finding the right investors can be difficult, because the innovation has to fit in the portfolio and strategy of the investor. Investors have strategies or guidelines on which companies and sectors that they will or will not invest in, for instance only in high-tech or agriculture, or in start-up companies that need less than $1 million or more mature companies that need more than £10 million. Moreover, their willingness to invest may be low in hard times. Professional advice for identifying good candidate investors and getting an introduction to the investors can significantly increase the chances of success.
Once the contact has been made, the convincing has to start with:

- A good, credible, and engaging story.
- A competent, entrepreneurial leader with a personal, strong belief in success.

If the investor happens to be not interested in the proposal, it is important to understand the reasons why. If the investor does not believe in the quality of the proposal, the product, or the people, then time for internal reflection should be taken for considering making adjustments. If the reasons are that the proposal does not fit in the portfolio or it loses to better competitive opportunities, approaching another investor may lead to the required success. But the reason could well be 'wrong timing'. Investors can be trend followers and at certain times it is 'in' to invest in high-tech, or social media, or health and at other times it is less popular. If you are in the hype, getting money is easy; beyond the hype, it can be better to wait for better times or accept that the opportunity is gone. It can also be that the budget for that segment in the portfolio of the investor has been spent already and then waiting and returning at a later stage may still be an option.

Once the investor has invested, it is important to build a stable, constructive relationship. The most important building block for a successful relationship is 'delivering as promised'. The second best one is to keep investors informed on progress, in particular, when setbacks occur. This information does not need to be on a day-to-day basis, but frequently enough to prevent a shock reaction from bad news. Investors are used to the delays and setbacks in innovation projects, but hate unpleasant surprises. If that happens, the investor might change into a tough player that may stop providing more money or force the company into making unpleasant changes. One of the first action items that will be considered if the company is underperforming is to oust the CEO. Investors are supporters with a bite. They prefer not to bite and share in the success of the enterprise, but if need be they will take action to protect their investments, even if that could be to the detriment of the innovator.

Roy Lipski, CEO of Oxford Catalysts, recalls the pressures and learning points from his experience when he had to raise a substantial amount of money in a few days to finance a very attractive business opportunity at a time that the company was still not out of the Valley.

Raising money under pressure

Well, we started our road show that Monday in 2008 when Lehman Brothers went bust. Funnily, it was not that difficult getting meetings with potential investors; they welcomed relief from staring at the sea of red on their screens. But then they just sat there miserably in the 'certainty' that the world was coming to an end. Not surprisingly, getting new investors on board proved impossible. Fortunately, we managed to scrape just enough from our existing investors. Then the largest investor in the raise got cold feet and threatened to pull out unless we could find a few more millions from others. It felt like some sort of Darwinian test: "I'm with you as long as you don't need me."

The clock was ticking and time had almost run out, but I persevered, drawing on every contact and idea I had. Finally, an acquaintance introduced me to someone who, with the aid of some creativity and persuasion, was willing to help. We were able to structure a financial instrument that satisfied the large investor that sufficient others were willing to share the risks, although no cash was actually changing hands. It could have so easily ended differently; it was a very narrow escape. It was a lesson in perseverance and keeping the faith.

Finding the right investors is a bit like match making. You are seeking a match between the risk appetite, investment size, time horizons, industry sector, and various other factors. Importantly for a start-up company, you are looking for investors who already share your broad vision of the future; if they do not, you will find it hard to enlighten them. Putting effort to identify good potential matches is as important as having a good story, and remember once you have good investors to always take good care of them.

You are probably going to have to meet with a lot of people. Getting a good advisor who can help narrow the search is worthwhile, and in this case made the difference. Do not assume you know enough to predict the outcome when it comes to the world of finance; keep doing the right things and trust that the results will follow.

Of course, good timing is crucial, no point trying to raise money in the midst of a financial meltdown...

7.3 Launching customer

For a start-up company, having a launching customer is one of the nicest things to have and an important KSF. There are a variety of definitions for launching customer, ranging from the 'first (significant) buyer' or 'first customer that takes part in the development of the product' to 'trophy customer, a big-name customer that creates interest'. All these types of launching customers can be important; any customer is important for the business and the first customer, even a small one, represents a milestone. But the key aspect of a launching customer for a start-up company is 'commitment before the launch'. A launching customer is important for three reasons:

1. Investors love launching customers. When customers sign up for the product, investors will follow. A launching customer provides evidence that the product makes sense and gives the investor confidence that his investments will be successful. If an innovation project is stuck in the Valley, finding a launching customer is a very attractive option for getting out of it. Having a Letter of Intent stating that a client will buy the product once it is ready and meets certain agreed standards provides VIP status in investor land for new companies.
2. A launching customer can give important information about the relevant product features and in this way improve the product and ensure that it is fit for purpose before the launch. Feedback on the product will also assist in adjusting the product launch and the business model. Few business models are spot-on in the design phase and knowledge on how customers will use and value the product can fine-tune the product offering, the supply chain, and the revenue model. A business model needs continuous refinement on the basis of feedback from the market.
3. A launching customer attracts other customers for the same logic as it attracts investors. It brings confidence that the product will work. And the more reputable the launching customer is, the more impact it has.

It is important to have the right launching customer; a customer with standing in the industry, a customer that is trustworthy and that represents the same values and qualities as the product. In certain industries such as the aircraft industry, product development will not proceed very far without an adequate number of customers and in that industry there can be competition for being the launching customer. Interest can be based on the opportunity of being first in the market, but also for steering the development of the airplane in a way that is optimal for the business model of the launching customer. For instance, Southwest Airlines used its status as launching customer to guide the development of Boeings' 737 Max. A key element in Southwest's business model is low operational cost, and as a consequence, their objective in the development was improved efficiency and productivity. This was combined with a minimum of changes compared to the existing model for optimizing the operational processes with the existing fleet. Fortunately, these objectives were in line with the objectives of Boeing.

This example can also serve to illustrate a potential negative side of close cooperation with a launching customer. If the development becomes too specific and creates a competitive advantage for the launching customer, the product may become less attractive for companies with a different business model and consequently for the market at large. Other disadvantages can result when the cooperation is not successful or leads to delays and a serious setback for the product development process can result if the cooperation is discontinued.

Sometimes governments are interested to be the launching customer in order to stimulate desirable innovations in certain selected areas. Well-known areas in which this may happen in many countries are renewable energy and sustainable mobility. Many companies in the area of solar energy, biofuels, electric vehicles, or traffic control benefit not only from subsidies but also from having local or national governments as early customers. Although governments may not have the highest convincing image for commercial customers, they can still be of significant value as an early customer.

7.4 Strategic partners

Strategic partners can be, and often are, essential assets for a start-up company. A strategic partner provides access to missing capabilities, parts of the value chain or market segments, and can provide a shield during hard times. Typically for a techno-starter, strategic partners are larger firms that supply capital and specific capabilities in return for licensing rights or a share in the company. The advantages of a strategic alliance include:

- Each partner brings in complementary capabilities.
- It expands the reach of both parties.
- Both parties become more robust because their resources are extended and diversified.

As with most long-term relations, it is not easy or straightforward to find the right partner and the right partnership arrangements. The partnership must be mutually beneficial in order to be sustainable. Therefore, the process for selecting strategic partners should be developed and executed carefully. The process usually goes through various discrete steps:

- A feasibility study to identify the major issues and ensure that the strategic objectives of the partners are aligned and can remain so in the long term.
- An assessment of the potential partner in respect of its strengths and weaknesses and whether the two companies would fit in the way they operate and their positioning in the market. Company culture, vision, and management style should not be sources of conflict or friction.
- Negotiations should not only cover the contractual obligations of both parties on the deliverables but also include agreement on how the partnership will operate, and under which conditions and how the partnership can be terminated.

Thoughts about termination should be an integral part of the negotiations because only 50% of the partnerships are considered a success and 10% fail.

Usually one or more of the strategic partners are much larger than the start-up company and this can be both an opportunity and a threat. An alliance with a big-name strategic partner can give the company a boost and much increased potential. For instance, Spotify, a music-subscription service provider, partnered with Facebook and found that the alliance boosted revenues because Spotify users that linked their account to their Facebook account were three times more likely to become paying customers than those that did not. A successful partnership with a big player is a major competitive advantage; it allows a small company to compete with the big ones. But if the big partner does not live up to the spirit or letter of the agreement, there may be little that the start-up can do.

Another element to consider is whether the key objective for the company is long-term independence or a friendly takeover. If independence is the objective, then giving shares to the partner may not be the best start of the partnership, in particular, if the partner wants a say in the way the company is run, for instance, with a member on the board.

Thus, it is important to take time and make sure that the partner has been chosen wisely. The right partner can do the company a lot of good; the wrong partner can be long-term trouble and can do severe damage to the company. An example of the danger of the wrong strategic partner is given by the story of Acorn that needed a strategic partner in order to survive.[1]

Acorn and Olivetti

Acorn computers was founded in 1978 by Chris Curry and Hermann Hauser, and they became Cambridge legends as they were making an estimated half of the computers used in UK schools by the late 1980s. Acorn invented the ARM (Advanced RISC Machine), and this attracted the attention of the electronics giant Olivetti who took over 25% of Acorn in 1985. Acorn's sales had struggled in the previous year, so they had to sell part of the company to survive. The deal was that Olivetti got Acorn shares at below market price in return for their selling Acorn computers through their much larger, global distribution network.

This never happened. A year later, Olivetti had not sold a single Acorn machine!! It turned out that the sales force of Olivetti had been told by the CEO to sell only IBM compatible products, which Acorn was not, so the result was hardly surprising.

Shortly thereafter, Olivetti exercised its right to take 75% of the Acorn shares and they also put in their own controlling management. This gradually shut the company down, completely destroying everything that was there.

If a company is not benefitting from the partnership, time and effort should be spent on changing or discontinuing it, even if the partnership does not cost much. If a poor partnership stands in the way of a good partnership, it is not the actual cost that should be considered, but the opportunity cost.

But failures of even promising partnerships should not take away the importance of the right partner as a KSF. An example of a very successful strategic partnership is provided by the merger of Oxford Catalysts and Velocys. Both companies were in the struggle of safely crossing the Valley, but jointly they became the leading technology company in the area of distributed conversion of natural gas, coal, and biomass to clean[2] diesel. Recalling the event a few years later, Pierre Jungels, chairman of the Oxford Catalysts Group, called the merger *"a marriage made in heaven."*

[1] http://elleeseymour.com/2012/02/24/how-olivetti-stitched-up-acorn.

[2] In this context, clean means sulfur free and reduced emissions of soot compared to traditional diesel.

Oxford Catalysts and Velocys

Oxford Catalysts started as a spin-off from the University of Oxford with a portfolio of patents resulting from 15 years of academic research on a specific class of catalysts. The company started to commercialize two catalysts, a novel one for which applications needed to be developed and one with superior performance for processes in the oil and gas industry.

Oxford Catalysts has a state-of-the-art laboratory in a business park in Oxfordshire, UK, and successfully developed a number of potential applications for their novel catalyst. However, the companies that were interested in the applications were also small start-ups with limited financial means and progress was slow.

The catalysts for the oil industry created significant interest in the industry, but building the supply chain took longer than expected and entering the market proved to be difficult in a conservative industry.

Velocys is a spin-off from Battelle, the world's largest nonprofit R&D organization, attempting to commercialize a microchannel reactor. Velocys has R&D facilities in Plain City, Ohio, USA. The technical development of the reactor progressed successfully, but the company had problems raising the money required for building the supply chain. The business model included the in-house manufacturing of the reactor because the specifications of the reactor were so tight that it was believed that it would be prudent not to rely on external parties.

Another problem was that the in-house-developed catalyst was not active enough to extract the best performance from the microchannel reactor. The catalyst from Oxford Catalysts proved to have the required activity and the two companies started to cooperate in the development of the reactor-catalyst combination.

The synergy between the novel catalyst and reactor was so obvious that the companies entered into discussions about a strategic partnership. To ensure that the business strategies would remain aligned long term, it was decided that the companies would merge. The merger took the form of a reverse takeover of Velocys by Oxford Catalysts. Another reason was that Battelle was not interested in financing the expansion plans of Velocys and preferred to sell their spin-off. They made the takeover possible by accepting payment for a large part in shares from Oxford Catalysts.

The benefits of this strategic partnership were many and significant.

- *The combined company can offer a complete product, an integrated optimized catalytic reactor, rather than components.*
- *Velocys became the leading technology company in the small-scale gas to liquids business.*

- *Velocys transferred its catalyst research to Oxford Catalysts and each company became a center of excellence for microchannel reactors and catalysts, respectively.*
- *Velocys adopted the less capital intensive commercialization approach already adopted by Oxford Catalysts by outsourcing the building of the supply chain.*
- *The new company could attract new partners that financed part of the development costs.*

Both companies may have crossed the Valley successfully, but the combination did it faster and more robustly, and has much more commercial potential.

7.5 Advisors

The importance of good advisors cannot easily be overvalued. There is empirical evidence that access to risk capital and to good advice are the two most important external factors for success in innovation. Money is generally recognized as a key factor, but good advice, particularly from people that have been there before, is also key. An innovator wanders into uncharted territory and thus any information about the unknown is useful.

At several points in the innovation process, it will be necessary for the entrepreneur to consult external advisors for making the right choices and for meeting regulatory and legal requirements. The advisors will need to be specialists in their field, but will also need to have experience with start-ups in the specific line of business. Good advice is worth its price, even for a start-up company with continuous liquidity concerns. The right counsel can make the difference between success and failure and then the price will always be right. Legal and patent attorneys may have special tariffs for start-ups or be willing to consider special delayed payment arrangements.

The board should be considered as a permanent advisor, and it is important that there is good and open communication between the executives and non-executives. It is obviously very important to have a good and balanced mix of expertise on the board and not only scientists or business developers.

It is equally important to have a good and trusted accountant. It is a source of sound financial advice, helpful in keeping the business a 'going concern', and will provide a sign of confidence to worried shareholders.

Points where expert advice is advisable or required are:

- *Concept Business Plan*

Outside assistance with developing or reviewing the business plan can assist in creating or selecting the right business model or adopting better assumptions on costs and market potential, and also serve as a reality check.

- *Formation of the Company*

It is essential that the right legal structure is selected and properly signed and sealed. Although many elements are standardized, the devil is in the detail and legal scrutiny is required.

- *Founder Agreement*

The agreement determines the allocation, transfer, and vesting of stock, but it can also include rules on shared goals and terms for leaving the company.

- *Fund Raising*

The higher the amount, the more important it is to have a trusted financial advisor who knows the right investors for the business and the right conditions. An introduction to an investor by a trusted advisor can make the difference.

- *Intellectual Property*

Advice from patent attorneys on the strategy for trademark and patent protection is an absolute must. Patent application should be done by an attorney that understands the business and the limitations of the start-up.

- *Employee Agreements*

Many start-up companies will try to limit the cash cost and compensate with shares, and expert advice is needed on such topics. Also the patent rights should be agreed; patents will be in the name of the inventors, but the employees should transfer their rights to the company.

- *Commercial Agreements*, such as supply contracts and product guarantees

An innovative product brings risks to the first users, and as a consequence special guarantees have to be provided for making these risks acceptable to the customers and special insurance may be required to make these guarantees financially manageable. This can be a complex matter with services and advice needed from specialized companies.[3]

- *Partnerships*

Contracts between partners can be complex, both to negotiate and in content. Assistance from experts is needed, also to make sure that the exit options are clear and manageable. It may also be required to assess the potential partners on the quality of their output and the commercial robustness, and the required due diligence investigations may have to be outsourced to an independent third party.

Getting advice is easy, but finding and recognizing the right advice is not that simple. Since so many agents are willing to give advice in the area of innovation, a degree

[3] A prime condition for making these risks acceptable is standardization in the production and the performance of the product with strict protocols throughout the whole organization and the supply chain.

of skepticism is warranted and in case of doubt it is sensible to get a second opinion. Good advice has to combine the right expertise with genuine insight and interest in the innovation and the business. For a start-up, the top advisors and consultancy firms may not always be the best option. Expensive, top quality advice for a relatively simple issue may not be the best use of money.

Another problem can be getting too much advice when the advisor does not only provide an answer to the question at hand, but also voluntary assists in identifying other issues that could benefit from sound advice, and it will not always be easy to understand which advice is needed and which one not. Too much of a good thing can become very expensive. It may be useful to pay per problem rather than by the hour or agree a cap on the cost.

7.6 Timing

The relevance of the right time for success in innovation is often not recognized, but innovations do fail by ignoring this issue. It tends to be difficult to convince innovators of the importance of the right time, as they are driven by their own momentum. But it is important to realize that the right time for the innovator may not be the right time for the customers or the investors. Critical timing moments in an innovation project are the times for asking for money and the timing of the launch, and the latter both with respect to whether the organization is prepared for it and whether the market is ready. Selecting the right window in time is a KSF. It is not so much a matter of waiting for the window of opportunity; it is one of being prepared when the window opens up.

It is said that investors and the markets are always interested in 'good products', but this is all in good time. The interest of investors and customers in new products is not necessarily fixed at a certain steady level but can go through ups and downs. Interest levels follow an irregular wavy pattern and that pattern is only partially open for rational analysis.

As we have seen in Section 7.1, investors will at certain times not be interested even in the finest proposals for a variety of reasons: the product does not fit the portfolio, the budget for the sector has already been spent, an investment in that sector went sour, the sector is no longer of interest. At another time, the proposal could have been a winner, but at the wrong time, no proposal is good enough.

For the markets similar open and closed spaces in time can occur. Technologies that initially may be seen as promising and leading may lose their magic and prime position. Nuclear energy is a well-known example of a 'product' that goes through a series of ups and downs. Genetic modification of vegetables has been accepted in certain parts of the world and not in others. Nanotechnology is a most promising technology with many breakthrough applications, but there are concerns about the health aspects of submicron particles and there is grassroots resistance that may slow down the progress of this technology.

It remains difficult to predict whether the market is ready for breakthrough innovative products. The mobile telephone now is a product the world cannot live without, but at the start, it had little appeal in the mass markets for quite a long time.

But sometimes it is easier to see that the market is ready for certain types of products. Good applications in the area of mobile communication are likely to be successful with the right implementation strategies. Clean, renewable energies and Internet applications are examples of 'safe' areas for good products at the right price. But even in such 'safe' areas the product still needs to have the right features and focus. Facebook only became a successful product when the target customer base was changed from restricted to open.

Another timing factor is that the new innovative technology should not be (too far) ahead of the supporting technologies that it needs for commercial success. A well-known example from history is the development of the aircraft industry. The Wright brothers took to the air in 1903, but the true commercial success of the industry had to wait till the 1930s for the development of stronger engines and materials. The electric car was popular over a hundred years ago, but it failed in the end because the battery was not up to standard and this may still be the case. In the 1990s, major efforts were spent on the development of the fly wheel in cars for storing the brake energy in support of the battery. However, its success was modest because the system only works properly with a Continuous Variable Transmission in the car and that technology was not sufficiently developed at the time. The time may be ripe now.

'A bridge too far' is one of the minor KUFs and this can happen to the best, as the quote[4] from Mark Zuckerberg, founder and CEO of Facebook, shows:

> *"The biggest mistake we made as a company was betting too much on HTML5,[5] because it's just not there yet. . . . We burned two years."*

In retrospect, Facebook should have concentrated on the so-called native applications for its mobile strategy.

Finally, the organization has to be ready at the time of launch. This sounds obvious, but it can be difficult to achieve, in particular for a high profile launch. Not only the company but the whole supply chain has to be ready for it and that can be difficult with a radically new product because it is a first time for all the parties in the supply chain. Building in flexibility for absorbing disturbances and mishaps in the supply chain should be part of the commercialization strategy.

For start-up companies, the launch of the new product may well be a gradual one because there is a transition period between the first sales and the time when the product is sold commercially against standardized conditions. This allows the start-up company to build up the required organization in line with demand and incorporate feedback from the first users.

[4] Forbes, 11.09.2012.

[5] HTML5 (*HyperText Markup Language 5*) is the latest, not yet finished, improved version of the HTML-standard. HTML provides functionality to a Website.

8 External Factors

8.1 Subsidy

For a start-up, subsidy can make the difference; it can enable the start of an innovation quest or serve as a lifeline that saves the idea from going down. Subsidy is an important supporter in the innovation arena, but certain subsidies can have unintended problematic side effects that should be recognized and considered when applying for the subsidy. The degree of drawback depends on the type of subsidy; in general, the more generic the subsidy, the less the disadvantages. The three main disadvantages are:

1. The costs in time and money for submitting and reporting can be considerable, certainly for a small start-up company.
2. It can enable an innovation that has no sound business plan to stay alive too long and the venture fails in the end after a lot of effort.
3. The business plan is adjusted to fit the requirements of the subsidy program.

The third drawback can be the most dangerous for a start-up because it is not readily recognized and can jeopardize the outcome of a good innovation. Many subsidies are linked to specific programs and the innovations have to meet specific conditions in order to be eligible. When the business plan is adjusted for meeting the specific conditions of the subsidy program, the focus of the innovation effort may switch from meeting the requirements of the customer to meeting the requirements of the subsidy and the deadlines of the monitoring and reporting system. In the research domain, adjusting the plan may not be too much of an issue and cutting a research program into the appropriate pieces that will fit into the various subsidy programs is an integral part of the effort. But in innovation, a change in the business plan for the sake of a subsidy can be problematic.

Another negative side effect of subsidies can be that they block good potential business opportunities or desirable developments. For example, the subsidy on green electricity can be so high that it is 'commercially impossible' to convert biomass into green liquid fuels. The problem with subsidies is that the decision taking is divorced from the risk and opportunity assessment. Decision makers in the subsidy programs select against other objectives and priorities than the risk-taking entrepreneur.

Certain types of subsidies have a tendency to be unreliable because they are likely to become victims of their own success. They can be stopped rather abruptly when the budget has been spent, usually well before the end of the planned period. For example, subsidies on green energy typically do not last the full period that is used in the investment plan for assessing profitability of the project.

Generic subsidies linked to the commercial performance of the product usually do not interfere with the way the business is run. Examples are tax advantages for research costs or on profits made from innovative products.

If a business plan needs adjustment for getting a subsidy or is not viable without it, the innovation project has an amber light and consideration should be given whether to proceed with the subsidy or not. If a subsidy fits the business plan with little associated costs and brings the reward more in line with the risks to be taken, then one should welcome this factor for success with open arms.

8.2 Standards

Standards are blockers as well as facilitators for innovation. In its simplest form the impact of standards can be formulated as below:

- Established standards block radical innovation.
- New standards create fertile platforms for innovations.

Technical standards became highly important during the Industrial Revolution with the need to make parts for equipment interchangeable, such as screws, nuts, pipe sizes, and electrical wiring. A well-known example of a high profile technical standard is the railroad track gauge that enables rail cars and locomotives to operate on the tracks of other railroad companies and across borders. Before standardization, it was necessary to unload and reload freight at every change of gauge. At one time there were more than 20 gauges in the United States.[1]

In general, it is very difficult to change standards; that is why they are standards. Standards can be seen as a set of best practices agreed by the various stakeholders to the benefit of all. Standards are very important for controlling and managing a wide range of issues, such as quality, health, safety, and compatibility.

But breakthrough innovations often have the problem that they do not fit existing standards and cannot comply or need to operate suboptimally in order to comply. This can block or delay the introduction and acceptance of the innovation. An example is given by Steven Bowns on the aerospace industry when he contrasts the progress in military aviation with, for instance, unmanned aerial vehicles and the civil aviation where introduction of innovations is difficult.[2]

> *"In contrast, the increasingly complex and burdensome regulatory standards for civil aircraft require that any new technology have a significant and proven track record before it can be adopted. In other words, it must not be new. Accordingly, the civil aerospace sector has been considered by many component and system manufacturers to be slow moving, where quality control and conformance to standards are paramount, with cost and innovation secondary."*

[1] Standards are from all times. The Romans already made a similar standard: the width of the track on roads so that army wagons could travel more easily throughout the empire.

[2] http://www.aero-mag.com/features/18/20125/1375/

In aviation, safety is so important that it seems unavoidable that changing standards is a slow process. Standards create hurdles for new entrants and consequently can be seen as being defenders of established positions.

On the other side, being able to set the new standard can create a significant competitive advantage, as illustrated by the commercialization of the rival video systems in the previous century between Sony, Philips, and JVC. Commercial success did not go to the technologically most advanced system, but to the system that was adopted as the new standard. For that reason, a company can aim to create a *'de facto'* standard for a product based on market dominance.

In the current postindustrial or digital era, the dominant areas of innovation are around ITC applications and the Internet. Setting standards in those domains can create significant competitive advantage and many a world-class company is striving to be a standard setter or at least be a major contributor to the standard setting effort. On the other side, many action groups and smaller companies argue that the standards for the Internet should be open because the Internet is a major platform for innovation with unprecedented potential, and standards should not be restrictive but enabling.

It is a rare occasion that a single techno-starter with a breakthrough invention creates a new standard as Edison did when he demonstrated his novel electric light system with a generator at 110 V AC that became the standard in the United States; simultaneously, he also standardized the screw base size and thread dimensions of electric lamp bulbs. This is quite an achievement, comparable to a solar energy company now persuading energy companies, governments, and the manufacturers of domestic appliances to adopt 16 volt DC as the new standard for electricity.[3]

In general, techno-starters and SMEs will avoid getting involved in standard setting but will base their ideas and concepts on the innovation platforms that are created by new or emerging standards. A potentially attractive way for an SME for getting involved in a standard setting process is by participating in a system innovation effort. In system innovation, a number of related innovations are developed in a coordinated way, and simultaneously, the required infrastructure and standards are designed and implemented. A simple example is the simultaneous development of battery cars and their loading points. Another example is creating a more sustainability society with the simultaneous development of renewable energies, recycling systems, changes in infrastructure, and possibly even changes in lifestyle. System innovation aims to create a more attractive innovation environment with lower risks for a group of innovators. System innovation tends to be nationwide and the government has a coordinating and stimulating role, but system innovation can also be carried out for a complex organization or a metropolis.

[3] If this could be done, the future for solar companies would look even brighter and bring additional safety for users and benefits for many developing countries.

8.3 Open innovation

Henry Chesbrough started his landmark book *Open Innovation*[4] with these sentences:

> *"Most innovations fail. And companies that don't innovate die. . . . Innovation is vital to sustain and advance companies' current businesses; it is critical to growing businesses. It is also a very difficult process to manage."*

Chesbrough argued that for most industries the traditional model where all the innovation effort of the company occurred inside was no longer optimal and should be replaced with a model where the internal effort is combined with external efforts in a collaborative mode. The external efforts can be as limited as idea generation with potential partners or customers, but more typically, the objective of open innovation is the use of external capabilities and creating access to the supply chain and markets.

The shift from closed to open innovation requires a change in mindset of management and staff and a different appreciation of what the winning factors are in innovation. Chesbrough drafted a list of contrasting principles between closed and open innovation, and a few examples are given in Table 8.1.

The shift from closed to open innovation is most clearly seen in large corporations that used to have a preference for a business model based on having all the knowledge that was needed for understanding and controlling the whole value chain in-house.

For start-ups, the shift in the mode of innovation is less fundamental. Small companies never had the luxury of having all the required capabilities in-house and always needed to work via partnerships or other cooperative arrangements. Furthermore, they typically lack the financial strength of large companies that can pay for outsourcing part or all of the innovation effort.

Open innovation can offer two advantages for start-ups:

- It can facilitate access to required expertise and capabilities. Since almost all companies these days are interested in cooperative innovation, and also universities and

Table 8.1 Contrasting principles in innovation

Closed innovation	Open innovation
• If we are the first to market, we will win • If we create the most and the best ideas in the industry, we will win • We should control our IP so that our competitors do not profit from our ideas	• Building a better business model is better than getting to market first • If we make the best use of internal and external ideas, we will win • We should profit from others' use of our IP, and we should buy others' IP whenever it advances our own business model

[4] H. Chesbrough, *Open Innovation: The New Imperative for Creating and Profiting from Technology*, Harvard Business School Press, 2003.

research institutes want to work on innovative efforts in a variety of arrangements, it has become easier for a small company to operate in the 'open mode' of innovation than it used to be.
- It can provide a source of income by carrying out development work that is outsourced by large companies. Although the latter mechanism carries a risk that the focus on own product development is eroded, it can help a start-up through the Valley or maintain its research capacity.

An organization needs to be prepared for open innovation in order to successfully integrate the internal and external knowledge and capabilities. All too often a company has no expertise or processes in place for the buy-in of external innovations or cooperatively sharing knowledge with third parties.

8.4 Sustainable innovation

For most techno-starters, the labels 'green' and 'sustainable' for their innovative product will be considered an asset, if not an essential requirement. Greenness and sustainability are valuable properties of an innovation; it would be an implausible proposition to develop a novel product that was unsustainable or ungreen. Sustainability and greenness offer a range of attractive platforms for innovation such as energy conservation, reducing harmful emissions, recycling, renewable energy, and most innovations in the 'using less' domain.

A well-known sustainability platform is Cradle to Cradle[5] (C2C) aiming to use all waste materials as a feedstock for another product and advocating a shift from a linear to a cyclical use of natural resources. A part of this approach is to upgrade recycling to upcycling. In recycling, the product becomes lower value in each cycle, whereas in upcycling, the recycling process creates a value product. The Cradle to Cradle Islands (C2CI) initiative in NW Europe has adopted this approach for making a number of islands self-sustaining in water and energy.

Another example is provided by the bioindustry that offers scope for a wide range of innovations for reducing the use of chemicals and fertilizers in agriculture or making the life of the animals more 'natural'. An example is FAB (Functional Agro-Biodiversity) in agriculture where cultivated fields get edges with flowers for attracting beneficial insects which has led to reduction of over 90% in the use of chemicals. The use of biomass for making chemicals, fuels, or electricity is also a source of creative innovation aimed at creating a bio-based economy with biorefineries as the core industry.

But it is useful to consider a few potential side effects of greenness that could be problematic. Greenness and sustainability have similar pros and cons as subsidies: it is

[5] M. Baumgarten and W. McDonough, *Cradle to Cradle—Remaking the Way We Make Things*, North Point Press, 2002.

a good thing to have them, but if the success of the innovation depends on it, the project has an amber light.

The first concern is that it can be open for debate whether a new product is green or not, and moreover, the opinion about this may change over time. For example, at the end of the previous century, it was considered green to use palm oil for substituting diesel fuel as a renewable fuel that would reduce CO_2 emissions. But the price of palm oil for use in cosmetics and food increased rapidly and palm oil became too expensive for energy use, and as a result, many small investors that had built conversion units went bankrupt. Furthermore, the conflict between food and energy made the idea unpopular and the whole concept collapsed. Some people see nuclear energy as a green option because of the lower CO_2 emissions; others see the radiation and radioactive waste as a threat to current and future generations.

Another concern is that scientific evidence on greenness and sustainability can also be open for reinterpretation over time. For example, the smog in Los Angeles was first attributed to CO, subsequently to SO_2, but in the end, ozone appeared to be the real culprit. Fortunately, the measures taken for reducing CO were good in their own right, but measures for reducing SO_2 had negative side effects because the NO_x levels in the air increased.

In the 1980s, acid rain was considered a very serious problem in Europe threatening to destroy the forests. The scare made for quick reduction in SO_2 and NO_x emissions, but the theory behind acid rain was only correct in a limited sense.

Finally, it can be difficult to check whether the product is still green when the whole value chain is taken into account. There are various ways[6] to carry out environmental impact calculations for the whole supply chain, but the results from different approaches can be conflicting. The outcome may still have great illustrative value but typically has too many arbitrary assumptions for serving as proof for greenness.

A specific form of 'sustainable innovation' is inclusive innovation for developing countries. Inclusive innovation aims to include the poor (the excluded) in the benefits of innovation. Pursuit of inclusive innovation today is considered not only essential but also inevitable for sustainable development.[7] The potential negative side effect for this form of sustainable innovation is that well-intended idealism rather than the real customer needs becomes the dominant driver. Inclusive innovation done from a laboratory or incubator in a developed country is bound to be unsuccessful because the understanding of the markets, customer preferences, and objections is inadequate. The experience with the export form of Appropriate Technology for Africa provides a clear example. In the previous century, there were various idealistic efforts to develop low-tech innovations for Africa that would fit the technical infrastructure in the countries and would be easy to use, maintain, and repair. The effort did not meet with great success because the customers

[6] Examples are VSM (Viable Systems Analysis), SVCA (Sustainable Value Chain Analysis), Carbon Footprint.

[7] International Conference on Creativity and Innovation at Grassroots (ICCIG, 2012).

considered it inferior or poor people's technology and were more interested in high-tech innovations adapted to the African environment.

Innovation for developing countries has to be done in the country itself and has to include local experts. The list of the 15 most important innovations[8] collected by the *Mail and Guardian* in South Africa provides an illustration that really relevant innovations are created within the country and that external technology and support can be important and relevant contributors.

Africa innovations: 15 ideas helping to transform a continent

1. *Hippo water roller*: The Hippo water roller is a drum that can be rolled on the ground, making it easier for those without access to taps to haul larger amounts of water faster.
2. *The iCow app*: Harnesses the power of cell phones to encourage best practice for dairy farmers and increase milk production.
3. *Farmer managed natural regeneration*: FMNR restores existing trees on drought-stricken land to improve Senegal's dwindling harvests.
4. *Portable water pumps*: Portable irrigation technology helping sub-Saharan smallholder farmers grow crops out of season.
5. *The Cardiopad*: A computer tablet diagnoses heart disease in rural households with limited access to medical services.
6. *Nigerian computer tablet*: The Inye computer tablet that can connect to the Internet via a dongle surmounts the price and infrastructure barriers in one go.
7. *Ethanol cooking oil plant*: Refining locally sourced cassava into ethanol fuel to provide cleaner cooking fuel.
8. *Refugees United*: Realizing the futile paper trail that many refugees were faced with when looking for missing relatives, the Danish brothers David and Christopher Mikkelsen wanted to find an easier way that refugees could trace their families.
9. *DIY aid supplies*: Aims to make Africa self-sufficient in emergency relief supplies.
10. *Sickle cell disease research*: Carry out scientific research on sickle cell disease and show that large-scale, cutting-edge genomic studies are possible in Africa.
11. *M-Pepea*: Offers emergency credit through cell phones to people who do not have access to credit cards or bank loans.
12. *The Tutu van*: The brightly colored 'Tutu Tester' van is a mobile clinic that incorporates screening for tuberculosis and HIV into a general health check-up in order to overcome the stigma associated with these diseases.

[8] http://m.mg.co.za/index.php?view=article&urlid=2012-08-26-africa-innovations-15-ideas-helping-to-transform-a-continent.

13. *Orange sweet potato*: Breeding sweet potatoes that contain beta-carotene to help in the fight against childhood blindness.
14. *Speaking Books*: A range of easy-to-use audio books designed to get potentially life-saving health messages out to millions of isolated people struggling with depression and mental health problems.
15. *Narrative Exposure Therapy*: Therapy for former child soldiers, encouraging storytelling to help them come to terms with their experiences.

Collaboration between developed and developing countries is important, and support to local innovation needs to be given by providing access to technology, risk capital, and business acumen. Emergent innovation[9] is an example of this type of sustainable innovation, whereby multinationals support local entrepreneurs in university incubators with money and business expertise in return for the 'first right of refusal'.

It is as important for sustainable innovation as for any innovation that the business case is sound; typically loss-making, idealistic innovations do not last long and in the end are thus not sustainable innovations. In the previous century, Philips successfully sold for many years a good cooking stove in India. Although it was a loss-making proposition, it was justified in that it was supporting the brand. However, when the overall business hit a dip, sales of the cooking pot were discontinued as one of the cost-reduction measures.

An interesting form of sustainable innovation is 'reverse innovation', whereby low-cost products are developed and tested at relatively low cost in emerging countries such as India and China and then introduced at low prices in rich countries. The low price opens up new markets for the product. An example of reverse innovation is provided by GE that is now selling an ultraportable electrocardiograph machine in the United States at an 80% markdown for similar products. The machine was originally built by GE Healthcare for doctors in India and China. India is now exporting its internal form of reverse innovation called 'frugal innovation'. Frugal innovation accepts limitations in resources, be it financial or material, as given and turns these constraints into advantages. *"Creativity loves constraints."*[10]

In summary, greenness and sustainability are powerful drivers for innovation, but knowing what green and sustainable is and will stay is not necessarily straightforward and open for diverging opinions. It is a major asset if the labels green and sustainable of a product are not in doubt and not subjected to challenge. Greenness should be an intrinsic aspect of a product just as quality and reliability. A product does not become a good product because it has a green label; a good product is green.

[9] J. Verloop, Emergent innovation, *EMPI Journal of Interdisciplinary Perspectives*, 2 (2), 2008.

[10] Quote from Marissa Meyer, CEO of Yahoo.

8.5 Innovation environment

Operating in a positive, supportive innovation environment is an important asset for an innovator and can make the difference between success and failure. Two aspects create a supportive environment for innovation: culture and infrastructure.

Innovation culture is a fairly vague concept with many different elements, such as support for entrepreneurship, admiration for risk taking, valuing education and knowledge, but the most relevant aspect could well be the response to failure. With failure intrinsically more frequent than success in innovation, a negative response to failure will create a major hurdle for subsequent efforts. In several European countries, failure in an enterprise can be 'a sentence for life', particularly in combination with bankruptcy. In contrast, in the United States, failure is seen as a painful, but valuable, learning exercise and an innovator that has failed before is valued as one with experience and perseverance. This difference in appreciation is important because there is empirical evidence that a failure does improve the chances of success for the next attempt.[11] Valuable experiences do not only come from successes but particularly from failures. Success often depends on specific circumstances, but failures tend to be generic. The importance of a positive attitude toward failure caused Paul Iske to start the Institute of Brilliant Failures.[12]

Institute of Brilliant Failures

The goal of the institute is to bring about a shift in the way failure is viewed by promoting a positive view of failure through the use of stories, films, interactive workshops, and road shows. A brilliant failure needs to fulfill four conditions:

- It targets on reaching a positive, valuable objective.
- Unnecessary mistakes have been avoided by proper preparations.
- The original objective was not achieved, but the resulting outcome had value.
- The failure brought learning or inspiration to others.

A well-known example of a brilliant failure presented by the institute was the discovery of America by Columbus:

- The original objective of Columbus was to find a shorter and safer trading route to The Indies.
- Columbus prepared his quest very well and was sponsored by the Queen of Spain, the crews were experienced, and the ships were well equipped.
- Columbus discovered beautiful, tropical islands, and thought that he had reached The Indies, but it turned out to be the Caribbean. Certainly a result with value.
- The discovery of America was an inspiration for many other discoverers.

[11] www.tweedekans.nl.

[12] www.briljantemislukkingen.nl.

The institute is also developing a checklist for organizations to assess their standing with respect to cultivating 'A Brilliant Failure Culture'. The checklist will measure three organizational development themes related to a brilliant failure attitude:

- Easing on the control button by management.
- Encouraging the right type of risk taking.
- Recognizing the value of, and learning from, failure.

Innovation infrastructure represents the physical aspects of the innovation environment. Innovation infrastructure covers a wide range of facilities and services including education and research facilities, legal and financial systems, transport and communication, incubators. All these infrastructural elements are supporting success in innovation and increase the potential for it. For start-ups, access to formalized know-how and risk capital are critical factors for success.

The presence of incubators can make a difference. An incubator can provide not only low-cost facilities but also access to services and advice in the areas of patent application, contacts with investors, development of business plans and company structure. Following the incubator stage, a move to a science park or similar center for start-ups is a logical and usually attractive next step. Many science parks are based on the grounds of research institutes or laboratories of major companies that have opened the gates and benefit of the lower costs and simpler execution of outsourcing and open innovation.

The value of science centers as breeding grounds for radical innovation may be overrated at times, because the exchange of information on the advanced technologies and innovations tends to be limited. But they are excellent platforms for networking on other problems associated with starting a business; for instance, exchanging information on good advisors for specific issues can be very valuable. The Cluster Innovation Centre,[13] the incubator at the University of New Delhi, has recognized the importance of a supportive ecosystem for innovation and is developing a comprehensive set of services that goes beyond the typical package of an incubator, but also includes developing innovative degree programs, training, and orientation modules on innovation, facilitating collaboration with industry and meaningful linkages with society. Networking and cooperation are important elements for success in innovation.

The relevance of the various infrastructural elements is straightforward, and the absence or the low quality of the innovation infrastructure in emergent economies is one of the hurdles for success in innovation. Innovation in developing countries tends to start with low-cost copying or modifying existing products from developed countries, and then can take off with an active Science and Technology (S&T) policy

[13] http://cic.du.ac.in/cic/.

from the government. This stimulates industrial innovation, but the importance of grassroots innovations in emerging economies should not be undervalued. The S&T policy in India may serve as an example of this duality. Ashok Jain[14] argues that the current split in the Indian policy, one for formal, industrial and one for rural grassroots innovation, is suboptimal because it denies grassroots innovators ready access to formalized research and knowledge.

8.6 Serendipity

Serendipity means 'happy accident', specifically the accident of finding a solution for a problem without looking for it. Many researchers and innovators are very grateful to serendipity for showing a way out when the effort seemed to be stuck. Many discoveries are linked to serendipity, such as the discovery by Pasteur of a vaccine against cholera, penicillin by Fleming, the benzene ring by Kekulé, vulcanization of rubber by Goodyear, Viagra by Pfizer. The microwave oven is a less well known example.[15]

The microwave oven

The microwave oven did not come about as a result of someone trying to find a better, faster way to cook. During World War II, two scientists invented the magnetron, a tube that produces microwaves. Installing magnetrons in Britain's radar system, the microwaves were able to spot enemy warplanes on their way to bomb the British Isles.

The idea of using microwave energy to cook food was accidentally discovered in 1945 by Percy Spencer of the Raytheon Company when he found that radar waves had melted a candy bar in his pocket. Experiments showed that microwave heating could raise the internal temperature of many foods far more rapidly than a conventional oven.

The first Raytheon commercial microwave oven was the 1161 Radarange, which was marketed in 1954. Rated at 1600 W, it was so large and expensive that it was practical only for restaurant and institutional use.

But serendipity typically is not a matter of luck or chance only. Knowledge and experience are needed to recognize the relevance and the value of the 'accidental finding', and the scientist or entrepreneur has to be totally involved in the problem and possess a deep understanding of the issues. There is an element of creating your own luck in

[14] Repositioning grassroots innovation in India's S&T policy: from divider to provider; *Current Science*, 103 (103), 2012.

[15] http://www.ideafinder.com/history/inventions/microwave.htm.

serendipity by focus, involvement, and awareness. As Pasteur once noticed (in slightly different French words): *"serendipity only happens to those that are prepared."*

Kekulé claimed that he had discovered the ring shape of the benzene molecule after having a day-dream of a snake seizing its own tail. This vision came to him after years of studying the nature of carbon-carbon bonds. The story of Pasteur as recorded at the History Learning Site[16] underlines the point that serendipity depends to a large degree on deep understanding.

> *"Pasteur knew about the work done by Edward Jenner regarding smallpox. Pasteur reasoned that if a vaccine could be found for smallpox, then a vaccine could be found for all diseases. Pasteur did not know how Jenner's vaccination worked so he had to proceed searching for a chicken cholera vaccine using a process of trial and error.*
>
> *In the summer of 1880, he found a vaccine by chance. Chamberland, one of his coworkers, had inoculated some chickens with chicken cholera germs from an old culture that had been around for some time. The chickens did not die. Pasteur asked Chamberland to repeat what he had done but with a fresh culture of chicken cholera germs. Pasteur reasoned that a new culture would provide more potent germs."*

Not only scientists benefit from serendipity, but also entrepreneurs do, in particular, innovative entrepreneurs with a deep understanding of the business. John Fentener van Vlissingen, founder and chairman of BCD Holdings, designed a brilliant business model after watching a daring event on television. On May 28, 1987, Mathias Rust, a German pilot, flew with a Cessna 172P from Berlin via Helsinki to Moscow and after circling a few times over the Red Square landed on a nearby bridge. For thousands of miles, Rust flew with a small plane so low that the Russian radar did not spot the plane. Fentener van Vlissingen recognized the value that 'flying under the radar' could have for an emerging business.[17]

Flying under the radar

The third-largest travel management company in the world began with a half-page memo. In 1987, John Fentener van Vlissingen wrote a brief message to his staff explaining how business travel would be a smart way to expand and diversify the holdings of his family-owned company. The key elements of the business opportunity were as follows:

- *Carve out a place in the service industry.*
- *Take advantage of consolidation opportunities in the fractured U.S. business travel market, where 33,000 agencies operated independently.*
- *Leverage technology to quickly emerge as a business travel leader.*

[16] http://www.historylearningsite.co.uk/louis_pasteur.htm.

[17] http://www.us.bcdtravelinmotion.com/index.php?article_id=93860.

That short memo was the first step toward what would become BCD Travel. Van Vlissingen foresaw the globalization of the travel sector, where businesses would gain efficiency and save money by relying on single, large agencies to manage entire travel programs. One travel management company, armed with the latest telecommunications and computer technology, could book a company's business trips to anywhere, from anywhere. It was a visionary idea.

Van Vlissingen outlined a strategy to buy small agencies in the U.S. Southeast to create a superregional company. "In the beginning, we did not want to be known as a buyer," *he recalled. The family called the project 'Flying under the radar'. The plan to create this major player began to take shape when van Vlissingen bought Atlanta-based WorldTravel Advisors. He then set about purchasing strategically aligned companies, among them a technology company in Dallas, a travel agency in Orlando, and an agency in Washington, DC.* "After only three years, we merged all the companies into one that, at that time, had total sales of US$350 million, and we immediately became No. 9 in the U.S." *The move surprised the competition.* "Everyone thought, 'Who is that?'" *he recalled.* "We were on our way."

There is a case to be made that serendipity is an integral part of success in innovation. Innovations go through dips and narrow escapes; the final product tends to be different from the original idea and for all these changes and challenges sound solutions have to be found. One needs a bit of good luck for success in innovation, deep understanding, and 24/7 dedication support serendipity.

Summary III

Understanding the innovation arena and the actors is one of the requirements for success. Factors for success are favorable circumstances and can be instruments, conditions, agents, contributors, or platforms. Unfortunately, KSFs do not only have a positive impact, but they also tend to have potential negative side effects.
The twelve KSFs that have been discussed are:

1. *Patents* — Patents serve to create a degree of exclusivity to a company in defense of higher margins. For a start-up, the prime reason for applying for a patent is to make the company owner of the IP because that is a critical requirement for investors.
2. *Investors* — Finding the right investors and keeping them happy are key challenges for a start-up company.
3. *Launching customer* — A launching customer can improve the product and enhance the confidence level of investors and customers, but the product improvements should not lead to an exclusive competitive advantage for the launching customer.
4. *Strategic partner* — A strategic partnership should create a win-win situation and alignment of the long-term objectives. The agreement should include possible exit options.
5. *Advisors* — Good advice has to combine the relevant expertise with an understanding of the innovation and the business strategy of the start-up.
6. *Timing* — Success with investors and launching the product depends very much on the right timing. Missing the 'window of opportunity' can be a cause of failure.
7. *Subsidy* — Receiving additional money is important for success, but it should not lead to inappropriate adjustments to the business model.
8. *Standards* — Standards represent the collective best practices of all stakeholders. Existing standards tend to block radical innovation; new standards can provide platforms for innovation.
9. *Open innovation* — Open innovation is an attractive option for a start-up company for creating access to missing capabilities.
10. *Sustainable innovation* — For most innovators, the labels green and sustainable for their product will be considered an asset, if not an essential requirement. However, if success depends on these labels, the project has an amber light.
11. *Innovation environment* — Two aspects create a supportive environment for innovation: culture and infrastructure. A positive attitude toward failure is an important element of a supportive innovation culture. Facilities such as

incubators that stimulate networking and cooperation, and provide access to good advice, know-how, and risk capital are important elements of an innovation infrastructure.

12. *Serendipity* — Many innovators are very grateful to serendipity for showing a way out when the effort seemed to be stuck. One needs a bit of good luck for success in innovation; deep understanding and 24/7 dedication support serendipity.

Appendix

Option Value

The simplest way to estimate the value of an innovation would be a Net Present Value (NPV) calculation with a discount factor that reflects the risk that the innovation project is deemed to have. For simple projects with a short development time, this approach for estimating the value might be adequate, but the method has its shortcomings. It treats all the cost and income figures with the same risks and thus implicitly assumes that nothing will or can be changed during the project. But there will be changes in the project and decisions can be made for reducing the risk of the innovation quest or adopting alternative business opportunities, and these factors should be taken into account.

A better way for estimating the value of an innovation is with the Option Value (OV) because innovation is creating an option for generating future cash flow. This method can incorporate the difference in risks for the various stages in the innovation project. The chance that the revenue streams will be realized is smaller than the costs of the development, and this will make the value of the innovation smaller but more realistic. On the other hand, the project can be stopped when the product does not meet expectations before major investments are made, and this option will increase the value of the project.

The OV reflects the fair value of the innovative idea at a specific stage of development. It is the value a rational investor would pay for the idea at that point in time.[1] Calculating the OV with the Black-Scholes formula is not possible because the data for estimating the parameters will not be known sufficiently precise to allow the calculation. But it is possible to estimate the OV with Monte-Carlo analyses of the various business options and their chance of success. At each gate, three or more decisions can be made: project is stopped, it is continued as planned, or alternative approaches are adopted. Typical alternative approaches are, for instance:

- The development is stopped and, if possible, the IP is sold.
- The performance of the product is different from that expected and a new market segment has to be adopted with different volume and margin expectations.
- The supply chain is not developed by the company as planned originally, but outsourced to a strategic partner.

[1] There will be a natural tendency in any organization to be optimistic about the chance of success and this will lead to an optimistic OV. A good health check on the resulting value is to ask whether the company would be prepared to pay that amount of money to a third party for buying the idea and the associate know-how.

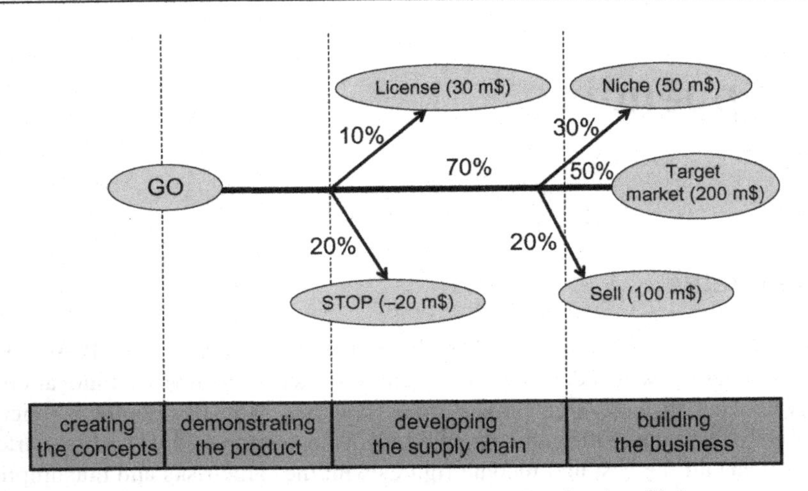

Figure A.1 Scenarios for estimating the Option Value.

Each business option is valued separately and is given a certain probability of success, and the Monte-Carlo analysis will give the average value, the extreme values, as well the distribution of values. If most of the outcomes are negative, the innovation may not be very attractive even if the average value is positive. This can happen when a few outcomes are very good, but most are negative. It is good practice to limit the number of scenarios to realistic options that have a fair chance of happening, for example, a probability of above 10%.

It is possible to take a shortcut approach, and rather than calculating a 1000 possible outcomes, only calculate the specific business options and derive in this way a Simplified Option Value (SOV). This approach gives less information but is simple and interpretation of the results is easier. Such an analysis is illustrated in Figure A.1 based on typical decisions that can be made at the tollgates about the progress of the project and the fate of the innovation in the market.

The Simplified Option Value

The example will be based on the business opportunities as used in Section 3.2 and illustrated in Figure A.1. The values for the various options are assumed to be STOP: −20 m$; license: 30 m$; target market: 200 m$; niche market: 50 m$; sell: 100 m$. The value of the STOP option is taken as equal to the development cost. The license value is taken as similar in size to the STOP value because that is a not untypical relation. The costs for developing the supply chain and the product launch are not explicit in this example, but they are included in the value of the final options.

Figure A.1 indicates the scenario chances that a specific option will be selected at a tollgate. The resulting SOV at the first tollgate is $0.2 \times -20 + 0.1 \times 30 + 0.7$

$(0.2 \times 100 + 0.3 \times 50 + 0.5 \times 200) = 94$ m$, and this could be considered to be the rational value of the company at that point. The extreme values are −20 and + 200 m$, with chances of occurring 20% and 35%, respectively.

At the second tollgate, when the product has been demonstrated and the decision to proceed has been taken, the SOV increases to 135 m$.

However, if the demonstration had been only partially successful and the option for the target market would have disappeared, the value of the sell option would reduce to, say, 50 m$, and the SOV would also reduce to 50 m$ (see Figure A.2). Such a result may lead to a different approach for the development of the supply chain. The option to license the technology could also be adopted. If this would be in addition to developing the niche market, the OV would increase; if it would be instead of, the OV would decrease.

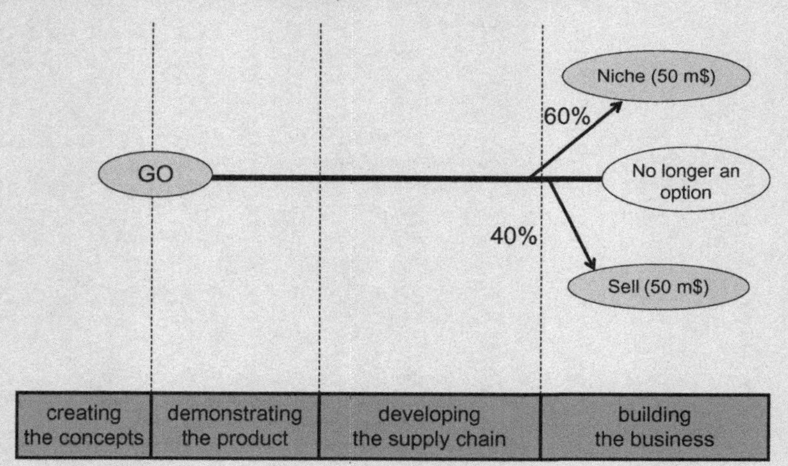

Figure A.2 Final scenarios for estimating the Option Value.

The value of this exercise is twofold. Firstly, it quantifies the many, often implicit assumptions made about the project and shows the impact of the associated risk on the value. Secondly, and possibly more important, it forces the innovation team to think the whole project through and seriously discuss and assess alternative outcomes. All too often, alternative outcomes are ignored and only the base case is analyzed and assessed in great detail. In the option analysis, the intrinsic uncertainty of an innovation project is explored and quantified. Detailed analyses are avoided because detailed data are bound to be wrong, and rough, realistic assumptions are appropriate.

The key requirement for the OV is that it should be positive, indicating that the chance of a positive outcome is better than a negative one. The OV should not be out of line with the cost for realizing the business. An OV of $100,000 for a million dollar business is attractive with a development and investment cost of $100,000, but doubtful if these costs would be a million dollars. The OV approach allows the innovator to track how the innovation project is developing and provides a handle to make early decisions to stop the innovation project in a transparent way. In a good development process, the OV will increase in steps every time a new assessment can use lower risk factors. However, in many developments, the OV will drop because further technical and market research shows that the market potential is less than expected initially. At the start of the innovation project, it may have been assumed that the novel product would have widespread or even global application, but in the course of development, it appears that the product will have to start in a niche. It can occur that with delays in the development and increases in the costs the OV becomes negative and it is time to stop the innovation project. This option for future cash flow will not be exercised.

Index

Note: Page numbers followed by *b* indicates boxes, *f* indicates figure, *t* indicates tables and *np* indicates footnotes.

Printed in the United States
By Bookmasters

Printed in the United States
By Bookmasters